Lecture Notes in Biomathematics

Managing Editor: S. Levin

27

Norman MacDonald

Time Lags in Biological Models

Springer-Verlag
Berlin Heidelberg New York 1978

Author
Norman MacDonald
Department of Natural Philosophy
The University of Glasgow
Glasgow G12 8QQ/Scotland

AMS Subject Classifications (1970): 92

ISBN-13: 978-3-540-09092-2 e-ISBN-13: 978-3-642-93107-9
DOI: 10.1007/978-3-642-93107-9

Softcover reprint of the hardcover 1st edition 1978

2141/3140-543210

PREFACE

In many biological models it is necessary to allow the rates of change of the variables to depend on the past history, rather than only the current values, of the variables. The models may require discrete lags, with the use of delay-differential equations, or distributed lags, with the use of integro-differential equations.

In these lecture notes I discuss the reasons for including lags, especially distributed lags, in biological models. These reasons may be inherent in the system studied, or may be the result of simplifying assumptions made in the model used. I examine some of the techniques available for studying the solution of the equations. A large proportion of the material presented relates to a special method that can be applied to a particular class of distributed lags. This method uses an extended set of ordinary differential equations. I examine the local stability of equilibrium points, and the existence and frequency of periodic solutions. I discuss the qualitative effects of lags, and how these differ according to the choice of discrete or distributed lag.

The models studied are drawn from the population dynamics of single species (logistic growth, the chemostat) and of interacting pairs of species (predation, mutualism), from cell population dynamics (haemopoiesis) and from biochemical kinetics (the Goodwin oscillator). The last chapter is devoted to a population model employing difference equations. All these models include non-linear terms.

The examples are largely new or drawn from my own papers, and the choice of mathematical methods, as well as of biological systems, reflects my interests. The main topic omitted, and it is one in which lags have been studied since the time of Lotka, is epidemiology. This has been surveyed in Volume 1 and 6 of this series, so that I only look at one recent paper, which happens to fit into my discussion of the logistic model Also models of predation are only discussed briefly, since a very detailed treatment is available in the lecture notes by J.M. Cushing, published as Volume 29 of this series.

The audience aimed at is primarily one of biologists with some familiarity with modelling in terms of ordinary differential equations, and interested in the possibility that lags may have significant effects in the systems they are studying.

Consequently rigour and generality have not been sought, and where recent mathematical results are presented these are ones that can be expressed as algorithms to be applied to the equations of the model.

While preparing this book I received many helpful comments and criticisms from colleagues and others. I wish to thank David Murray-Smith, Sandy Watt and Tom Wheldon in Glasgow, Bob Mohr and Bob Rosen at Dalhousie, and John Beddington, Kenneth Cooke, Zvi Grossman, Mike Mackey, Robert May, Paul Rapp and John Tyson elsewhere. I also wish to thank some of those named, as well as D.J. Allwright, J.M. Cushing, H.C. Morris and U. an der Heiden for sending material in advance of publication. Margaret MacDonald is thanked for valuable advice on style and Kay Johnston for preparing the final typescript. Finally I wish to thank J.C. Gunn for encouraging my transformation from a nuclear physicist to a biomathematician, and the Court of the University of Glasgow for allowing me a year of study leave, during which this book was written.

CONTENTS

FIGURES

INTRODUCTION

1a Discrete and Distributed Lag

One of the most popular styles of modelling in mathematical biology makes use of first order differential equations. The rates of change of the variables are taken to depend only on their current values, as in classical dynamics. This implies that the previous history of the variables, as well as their future behaviour, can be obtained by integrating the equations. However in biological models it is often more appropriate to allow the rates of change of the variables to depend on the previous history. Thus prediction requires the specification not merely of initial values but of values over an extended time, and retrodiction by changing t to -t is meaningless.

This book is concerned with models in which at least one variable changes at a rate dependent on its past history, with some of the methods available for obtaining solutions of the appropriate equations, and with the effects produced. Some points should be noted about restrictions in the subject matter. The restriction to differential (or to difference) equations means that I am only concerned with deterministic models. The restriction to first order equations is no real restriction, since any higher order equation can be rewritten as a set of first order equations. The restriction to ordinary differential equations means that the systems studied are treated as if homogeneous in space.

In order to fix the terminology used, consider an equation of the form

$$\frac{dx}{dt} = f(x, z). \qquad (1)$$

Setting $z = x$ gives an <u>instantaneous</u> equation, in which the rate of change of x depends only on the current value,

$$\frac{dx}{dt} = f(x, x) = g(x). \qquad (2)$$

To obtain a particular solution of equation (2) for $t > 0$ one specifies the initial value $x(0)$. Setting $z = x(t-T)$ introduces a discrete delay or <u>discrete lag</u>, the equation (1) becoming a <u>delay-differential</u> equation,

$$\frac{dx}{dt} = f(x(t), x(t-T)). \tag{3}$$

To obtain a particular solution of equation (3) for $t > 0$, one specifies values of $x(t)$ over the range $-T$ to 0 of t. Setting

$$z = \int_{-\infty}^{t} x(\tau)\, G(t-\tau)d\tau$$

introduces a distributed lag, equation (1) becoming an integro-differential equation,

$$\frac{dx}{dt} = f(x(t), \int_{-\infty}^{t} x(\tau)\, G(t-\tau)d\tau). \tag{4}$$

To obtain a particular solution of equation (4) for $t > 0$, one specifies values of $x(t)$ over all negative t. I shall refer to the function $G(u)$ as the memory function. This somewhat anthropomorphic term is used in chemical physics for an analogous function. An alternative term is delay kernel. The memory function will be taken throughout this book to be normalised,

$$\int_{0}^{\infty} G(u)du = 1. \tag{5}$$

In this way one ensures that for (4) as well as for (3) the equilibrium point of the instantaneous equation,

$$x = x^{o}, \quad g(x^{o}) = 0,$$

remains an equilibrium point in the presence of lag, with

$$z = z^{o} = x^{o}, \quad f(x^{o}, z^{o}) = 0.$$

Equation (4) reduces to equation (3) if G is a delta function,

$$G(t-\tau) = \delta(t-\tau-T).$$

At various points I shall compare results for discrete and distributed lag. This will always be done in terms of the average lag $\bar{T}$ for the given memory function,

$$\bar{T} = \int_{0}^{\infty} uG(u)du. \tag{6}$$

In the physical sciences, although differential equations are the standard form for dynamical models, there are at least two traditions on which one can call for insight into delay problems. One is electronic engineering and control theory, in which the emphasis is overwhelmingly on discrete lag. This corresponds, for example, to the discrete time of propagation of a signal along a cable. There is an extensive literature on delay-differential equations, of which the books by Bellman and Cooke (1963) and by El'sgol'ts and Norkin (1973) are noteworthy.

The other tradition is the study of the elastic and magnetic properties of materials, in which it is frequently necessary to take account of how a specimen was prepared before one can predict its behaviour. Here integro-differential equations can be used, for example with the simple memory function

$$G(u) = \exp(-au).$$

When such equations were first employed, there was some dispute on the propriety of explicitly including the history of the specimen. One viewpoint was that this is made necessary by the existence of hidden variables. It was argued that if one had access to these variables one would be able to substitute for an integro-differential equation an extended set of differential equations. Working with the obvious variables one needs to specify delay explicitly, but with a complete set of variables this would not be necessary. Although this may seem a rather meta-physical point, it has an echo in the motivation for introducing delay in certain biological models. For an account of delay equations from this point of view one may consult the book by Vogel (1965). A general account of functional differential equations, which include both types (3) and (4), is to be found in the recent book by Hale (1977).

The idea of replacing delay by an extension of the set of variables lies behind the policy, emphasised by Vogel in particular but with quite a long history in the context of elasticity, of setting down conditions under which a set of equations with delays is equivalent to a dynamical system, that is to say to a set of differential equations. With a special choice of memory function this approach gives a convenient method, employed throughout this book, which I shall call the <u>linear chain trick</u>. Using

$$G(u) = G_a^p(u) = \frac{a^{p+1}\, u^p}{p!}\, \exp(-au), \tag{7}$$

equation (1) is supplemented by the equations

$$\frac{dz}{dt} = a(y-z),$$

$$\cdot \; \cdot \; \cdot \; \cdot \; \cdot \; \cdot \; \cdot \qquad (8)$$

$$\frac{dv}{dt} = a(x-v).$$

with p+1 extra equations in all. These extra equations are linear, and each links two successive members of a chain of variables starting with z and ending with x. I shall call p the order of the lag. It has to be emphasised that, for the memory function (7), to solve (4) from t = 0, with a given previous set of values of x(t) over the interval $(-\infty,0)$, is equivalent to solving the set of equations (1), (8) from t = 0, with an appropriately inter-related set of initial values x(0), ... z(0).

This method has been used, for example by Forrester (1961), as a convenient way of dealing with lags in computer simulations set up in terms of sets of ordinary differential equations. It can also lead to some analytical insight into the nature of lag effects, as I hope to demonstrate. It has recently been used in a number of investigations of biological models, in the special forms with order 0 or 1. Sometimes this has been done explicitly as a means of handling the memory functions $G_a^0(u)$ or $G_a^1(u)$, but in other cases it has been justified by the intuitive idea that somehow the effects of a lag ought to resemble those of an extra stage in the process examined. These references are listed in the supplementary bibliography, unless cited in a particular context in what follows.

1b Origin of Lags in Biological Models

I shall make a distinction between lags that are inherent in the system studied, and lags that enter a model because of the specific approximations involved in setting up the model. Both discrete and distributed lag may be appropriate. In physiology, systems are frequently modelled in terms of control theory. A feedback signal may be transmitted as a nerve impulse, so that the time of transmission introduces a lag of the order of a fraction of a second. Or the signal may be transmitted by the circulation of a hormone, with rather longer lag. Discrete lag may be an excellent approximation, but there is bound to be some spread of the lag around the mean value. Here the use of distributed lag may be seen as a way of allowing for a stochastic element in this aspect of what is otherwise a deterministic model. I shall expand on this in section 1f.

At the other extreme, there is a famous example in which discrete lag is totally inappropriate. This is to be found in the classical model of Volterra for predation in mammals or fish. In its simplest form this model consists of two coupled equations for the rates of change of the populations of the prey species, x(t), and of the predator species, y(t),

$$\frac{dx}{dt} = \varepsilon x - \alpha xy,$$
$$\frac{dy}{dt} = -\gamma y + \beta xy. \qquad (9)$$

Volterra was well aware that this model contains gross simplifications, one of which is that the equations (9) are instantaneous. In prey-predator encounters there is an asymmetry between prey and predator not reflected by these equations. The model assumes that an encounter between predator and prey can bring about an instantaneous change in the prey population, which is plausible, and also an instantaneous change in the predator population, which is certainly not. In fact one should describe the birth and death rates of predators in a manner that takes into account the abundance of prey over some previous span of time. In the case of the birth rate this would have to be at least the period of gestation. Volterra (1927) studied a modified model in which the second of these equations is replaced by an integro-differential one,

$$\frac{dy}{dt} = -\gamma y + \beta y \int_{-\infty}^{t} x(\tau)\, G(t-\tau)d\tau. \qquad (10)$$

He obtained results for general G(u), but one could plausibly use either $G_a^0(u)$, with mean lag $\bar{T} = 1/a$, or a square memory function,

$$G(u) = 1/T, \quad u \leqslant T,$$
$$G(u) = 0, \quad u > T,$$

with mean lag $\bar{T} = T/2$. Another example due to Volterra, in which discrete lag is manifestly absurd, will be described in section 4b, while predation models of this general type will be discussed in Chapter 7.

Typically in population dynamics one may expect lags that are intermediate between this type and the discrete type, so that the memory function has a peak near

$\bar{T}$. This is because a natural delay time often corresponds to a generation span. Consider for example obligate mutualism, in which each of two species depends on the other. Plants of a certain species are pollinated by insects of one species only, and these obtain nectar to feed their larvae from this one plant species. It is natural, as a first approach to a model of this interaction, to work in terms of the number of adult insects and mature (flower-bearing) plants. One then has to recognise that successful pollination is reflected in the number of new mature plants after some interval of time. An adequate supply of nectar is reflected in the emergence of a new generation of adult insects from the larvae, again after an interval. A discrete lag may be too crude an approximation, a memory function such as $G_a^p(u)$, with $p \geq 1$, being preferable. It may be noted that in this example the distinction between lag intrinsic to the biology, and lag introduced because of the nature of the model, becomes blurred, since I have specified that one constructs the model in terms of adults only. This topic will be resumed in the next section.

It is rather unlikely that in any biological context one has data that specifies $G(u)$ precisely. One advantage of the form $G_a^p(u)$, as will be seen later, is that one can sometimes set down results for general p. These will include results for discrete lag as a limiting case, obtained by letting p and a increase indefinitely, while keeping the ratio $\bar{T} = (p+1)/a$ fixed.

In population models it may well become necessary to have two successive lags, such as a generation lag coming after a lag of the kind considered in the Volterra model. Thus an insect predator survives and is fertile by virtue of the prey available during its adult life, lays its eggs and dies. After some lapse of time new adults appear. This involves convoluting two distributed lags, so that the variable z is defined by

$$z = \int_{-\infty}^{t} G_2(t-\tau_2)\, d\tau_2 \int_{-\infty}^{\tau_2} x(\tau_1)\, G_1(\tau_2-\tau_1)\, d\tau_1 . \tag{11}$$

With $G_1(u) = G_a^p(u)$ and $G_2 = G_a^q(u)$ the convolution is trivial; this will be discussed in 2d.

1c Lag as an Alternative to Age Structure

Consider for example the insect member of the mutualistic system of 1b. It may not be necessary to model a generation span as a delay if, instead of using adult populations only, one sets up a more detailed model in which the population dynamics of eggs, larvae, pre-adult instars and adults are all explicitly specified. One may of course not have data complete enough for this to be possible. This argument parallels the comment made above about hidden variables. I shall illustrate this further in a rather different context, that of haemopoiesis, the manufacture of blood cells in the bone marrow.

It is generally agreed that bone marrow contains proliferating pluripotent stem cells, capable of differentiating into any one of a number of lineages of developing cells which culminate in the various forms of mature blood cell. In the case of the type of white cell known as granulocytes, this maturing process takes about a week or ten days. In various models, such as that of Wheldon et al. (1974) it is assumed that initiation of the granulocyte lineage is inhibited by some sort of chemical signal from mature granulocytes while they are still in the marrow, before exit to the blood stream.

One could set up two kinds of model of this control process. In the more detailed kind the variables x_1, . . . x_n are the concentrations of stem cells, first recognisable committed precursors of granulocytes, ... and so on until one reaches the mature granulocytes. The dynamics of each transformation are expressed in equations, and the feedback will appear by way of x_n-dependent terms in the rate equations of the stem cell and of the first precursor. All equations are instantaneous, and the lapse of time in the maturing process is occupied by the successive rise and fall of x_2, x_3, . . . x_n, after an initial pulse in x_1. However one does not know enough about the dynamics to formulate all these equations; also the only essential feature may be the feedback term. So one models in fact in terms of x_1 and x_n alone, or even only of x_n, assuming a constant self-replenishing reservoir of stem cells, as is reasonable in normal adult life. In this case the feedback must be given a delayed form. Such models will be examined in Chapter 6.

So the choice in a case like this may be between a lag and the introduction of age structure. In some cases the linear chain trick is formally identical with the introduction of intermediate stages which obey the simplest possible dynamics. An example is discussed in Chapter 5. However this appealing analogy turns out not to be valid in the models of haemopoiesis considered in Chapter 6. The "extra stages" formally added in this way in fact come after the mature cell stage.

1d Lag as an Alternative to Spatial Structure

Another reason for including a lag in a model may be that the real system has a spatial structure so that different stages in the process modelled take place in different regions. Then at least one aspect of the spatial structure, the finite time for an intermediate component to cross from region to region, may be qualitatively modelled.

Consider for example models of oscillatory biochemical processes in a cell, such as the models of glycolysis described by Goldbeter and Nicolis (1976) or the more abstract model of Goodwin (1965), discussed in Chapter 5. These models are set up in terms of ordinary differential equations, treating the cell as spatially homogeneous, with all the reactions taking place uniformly throughout the cell. The reactions may have a natural source of lag, such as the time taken to transcribe RNA. But they may also involve transport from nucleus or membrane to cytoplasm. The finite time of transport could be included as a lag, and one may wish to seek the most appropriate way to do this.

It will be seen in Chapter 5 that in the Goodwin model, the linear chain trick allows a rather elegant way of dealing with lag. The model already contains a chain of linear equations linking successive members of a set of chemical concentrations, and the lag is handled merely by lengthening this chain.

The equations (8) for the linear chain express an "active" or "one-way" process, rather like the propagation of a wave pulse. Discrete lag would be appropriate for a delay caused by a wave pulse propagating from one plane to a parallel one, or from the centre of a sphere to its surface. Linear chain lag would be more appropriate for pulses propagating from one finite region to another. Diffusion, however, is a "passive" process, like Lenin's "two steps forward, one step back". To simulate diffusion with a set of intermediate variables would require linear equations of the form

$$\frac{dx_r}{dt} = (a\ x_{r-1} - 2x_r + x_{r+1}), \qquad (12)$$

rather than of the form (8). Further thought makes it apparent that the interaction of two compartments A, B by way of diffusion through a line of intermediate compartments can be simulated by introducing a lag in the interaction of A and B and also delayed self-interactions of A and of B. This approach rapidly becomes more cumbersome than the diffusion problem itself.

1e The Effects of Lag

Turning from the reasons for including lag in biological models to the effects produced, both qualitative and quantitative questions arise. For any dynamical system there is a natural sequence of qualitative questions. One may ask how many equilibrium points are there, and are they stable or unstable against small displacements? For each unstable equilibrium point, where do the trajectories from slightly disturbed points end up? Do they lead to stable equilibrium points or to infinity? If these trajectories stay within a bounded region, do they tend towards a closed periodic trajectory or to some more complicated orbit, such as a strange attractor? This last possibility allows, loosely speaking, the filling up of a region of phase space. One may wish to know whether introducing lag alters the answer to any of these questions.

If a system is at equilibrium it can mean nothing to assert that the previous values of the variables matter, since these values are constant. So the model with lag must be set up so that all the equilibrium points of the original instantaneous model remain fixed, and there are no new ones. For distributed lag this is ensured by the normalisation of the memory function according to equation (5). An exception to this general requirement is mentioned in 6c.

It is natural to require that as the lag is made shorter, the stability or instability of an equilibrium point should be as in the instantaneous model. El'sgol'ts and Norkin (1973) present, for discrete lags, a proof that this is the case, so long as lags are introduced only in the variables and not in the derivatives in the set of first order differential equations. In the usual terminology, this desirable result holds for <u>retarded</u> equations, but may not hold for <u>neutral</u> equations. Neutral equations have rarely been used in biological models, although their use has been advocated by Lewis (1972) in the context of population dynamics. I shall return to this question, for the memory functions $G_a^p(u)$, in 2d.

In the models examined in this book, the most typical effect of lag is to destabilise an equilibrium point as the discrete lag T or the mean lag $\bar{T}$ is raised. In some cases, as discussed in Chapters 4 and 5, it can be proved that whenever the equilibrium point is unstable there is a non-constant periodic solution. In other cases numerical integration of the equations provides evidence of the periodic nature of the solutions.

The possibility of strange attractors arises when the number of first order differential equations exceeds two. It is possible that a model may acquire such a solution when its dimensionality is raised from two by the incorporation of the linear chain, or when a discrete lag is incorporated [Glass and Mackey (1977, 1978)]. Little is known as yet about such space-filling behaviour in differential equation models in biology. However a great deal of interest has been aroused by the analogous phenomenon of "chaotic" solutions of difference equation models as reviewed by May (1976b). I shall discuss a stabilising lag effect in this context in Chapter 8.

There are two obvious quantitative questions concerning lag effects. What are typical values of T or $\bar{T}$ that give a significant effect, such as destabilising an equilibrium point? When lag brings about a periodic solution of period T_{osc}, what is the ratio of T_{osc} to T or $\bar{T}$? Other questions may be of interest, when one can carry out both discrete and distributed lag calculations. Given that discrete lag with the value T destabilises an equilibrium point, and that so does distributed lag with mean lag $\bar{T}$, which of these times will be the smaller? Also given that discrete lag can destabilise an equilibrium point, is there a reasonable memory function such that distributed lag can not have this effect, however large $\bar{T}$? It will be demonstrated in this book that in the sense implied by these questions, distributed lag is less destabilising than discrete lag in a number of specific cases.

It is clear that in any instantaneous model certain parameters will express the natural time scales for the dynamics. For example in a prey-predator model such as that of Volterra, one such time scale is given by the doubling time of the prey population in the absence of predators. Another is given by the period of the population oscillations, if these occur. May (1973) has emphasised that for a significant lag effect to occur the mean lag should be comparable to the greatest of these natural times. He cites the case of a system of three populations, of plants, herbivores and carnivores, with a lag in the plant-herbivore interaction. This lag is taken to have mean value in excess of the doubling time of the uncropped plant population. The plant-herbivore subsystem has a stable equilibrium in the instantaneous case, destabilised by this lag. The herbivore-carnivore subsystem is taken to have stable oscillations with period greater than the mean value of this lag. The full system is found once more to have a stable equilibrium point.

I shall present in Chapter 4 detailed results for the ratio T_{osc}/T and the ratio $T_{osc}/\bar{T}$ in the logistic model for the growth of the population of a single species. In Chapter 5 I shall present results for these ratios in another model with a single non-linear equation. Some values of these ratios found in certain other models, by numerical integration of the equations, will also be cited. The various values span the range from 2 to about 5.5.

1f <u>Lags and Stochastic Models</u>

It is possible to make more explicit the argument at the beginning of 1b on the superiority of distributed lag to discrete lag. Let some process in a population model involve, for the individual member of the population, a discrete lag. (This is quite different from the situation in the predation model discussed in 1b, where the previous history is needed for each member of the population.) For example each member of the population may hatch from an egg laid a time T_i earlier. Over the whole population these times T_i are distributed in a random manner, with a probability distribution P(T) which is peaked near the average lag $\bar{T}$. If this peak is sharp enough $\bar{T}$ may be used as the unique lag in a discrete lag model of the population. Otherwise one should use a distributed lag model, the memory function being G(u) = P(u). Lewis (1977) discusses various probability distributions that may be appropriate in this context.

To go beyond this qualitative argument, it is necessary to say a little about the construction of stochastic models. Corresponding to a given instantaneous deterministic model for a variable x, it is possible to set up a linear stochastic model for the probability distribution P(x), which can be consistently expanded in terms of an appropriate small parameter. This small parameter can be a measure of the reciprocal of the system size, as discussed by van Kampen (1976). Or it may be a measure of the time scale of fluctuations in x as compared with a natural time scale of the deterministic model, as discussed by White (1977). Take as an example the logistic equation

$$\frac{dx}{dt} = rx(1 - x/K), \tag{13}$$

to be dealt with at length in Chapter 4. Here the population rises to a maximum value K, the rate of growth being on a time scale 1/r. In van Kampen's method the small parameter would be 1/K. White would postulate that fluctuations are fast enough for the deviations of x from its mean value to be uncorrelated outside some interval s, and his small parameter would be sr.

To lowest order this expansion yields two equations. One is a differential equation for the mean value $\bar{x}$. The stochastic model is taken to be an appropriate analogue of the original deterministic model by arranging that the equation satisfied by $\bar{x}$ is that satisfied by x in the deterministic model. The second equation is one for small Gaussian fluctuations about $\bar{x}$.

White has examined the analogous formulation for a deterministic model in the form of a delay-differential equation (discrete lag model), in which the fluctuations originate in variation in the lag. He shows that to this order of approximation $\bar{x}$ satisfies, not the original equation, but an integro-differential one for which the memory function is the probability distribution for the lag.

STABILITY ANALYSIS

2a The Linear Chain Trick

The stability analysis of equilibrium points, in the presence of distributed lag with memory function $G_a^p(u)$, can be carried out without interpreting the lag in terms of additional linear equations. However I believe that some insight into the results can be gained by keeping this interpretation in mind, so I shall begin by deriving the linear chain result. Starting with an instantaneous equation for a set of variables $x_i(t)$,

$$\frac{dx_i}{dt} = g_i(x_1, \dots x_n), \quad i = 1, \dots n, \tag{1}$$

I introduce a distributed lag in one term of one of these equations, so that the k'th equation becomes

$$\frac{dx_k}{dt} = \hat{g}_k(x_1, \dots x_m, \bar{x}_m, \dots x_n), \tag{2}$$

where

$$\bar{x}_m = \int_{-\infty}^{t} x_m(\tau)\, G(t-\tau)\, d\tau, \tag{3}$$

and

$$G(u) = G_a^p(u) = \frac{a^{p+1}\, u^p}{p!} \exp(-au), \tag{4}$$

where a is a positive number and p is a positive integer or zero. Then it is clear that

$$\text{(a)} \quad G_a^p(\infty) = 0,$$

$$\text{(b)} \quad G_a^p(0) = 0, \quad p \neq 0,$$

$$\text{(c)} \quad G_a^o(0) = a.$$

Renaming $\bar{x}_m$ as x_{n+p+1}, and defining new variables x_{n+1} to x_{n+p} as

$$x_{n+j} = \int_{-\infty}^{t} x_m(\tau)\, G_a^{j-1}(t-\tau)\, d\tau,$$

the results (a, b, c) imply that the new variables satisfy a sequence of linear ordinary differential equations

$$\frac{dx_{n+j}}{dt} = a(x_{n+j-1} - x_{n+j}), \quad j = 2, \dots p + 1,$$

$$\frac{dx_{n+1}}{dt} = a(x_m - x_{n+1}). \tag{5}$$

Supplementing equations (1) and (2) by the set (5) is equivalent to using (3) and (4), provided that the new variables are given appropriate initial values.

In most of the applications of this trick which I shall examine, it is sufficient to replace one of the variables in part of one of the equations, as here. However, one may wish to include an integral over past time of some function of one or more of the variables. As an example, consider the equation

$$\frac{dx}{dt} = xy. \tag{6}$$

According to the requirements of the model one might wish to replace x by

$$\int_{-\infty}^{t} x(\tau)\, G(t-\tau)\, d\tau,$$

y by

$$\int_{-\infty}^{t} y(\tau)\, G(t-\tau)\, d\tau,$$

or xy by

$$\int_{-\infty}^{t} x(\tau)\, y(\tau)\, G(t-\tau)\, d\tau.$$

These may be called <u>diagonal</u>, <u>off-diagonal</u>, and <u>total</u> lag respectively. For the first two cases the linear chain is obtained, with the right hand side of the final equation of the chain being $a(x-z)$ or $a(y-z)$ as appropriate. For the third case one has the equations

$$\frac{dx}{dt} = v,$$

$$\frac{dv}{dt} = a(w-v), \tag{7}$$

$$\cdots\cdots\cdots$$

$$\frac{dz}{dt} = a(xy-z).$$

the last one being non-linear.

Vogel (1965) and Fargue (1973) give a general condition to be satisfied by the memory function $G(t-\tau)$ so that the integro-differential equation can be replaced by a set of ordinary differential equations. This is

$$\frac{d^r}{dt^r} G(t-\tau) = \sum_{s=0}^{r-1} \alpha_s \frac{d^s}{dt^s} G(t-\tau). \tag{8}$$

which implies, in general, sums of G_a^p. Fargue also states a condition under which the integro-differential equation may be replaced by partial differential equations.

There are obvious advantages in making a uniform choice of memory function in a variety of problems, allowing one to alter the width and the position of the peak of the lag distribution. The form G_a^p is convenient because of the specially simple form of the new equations, and because it goes over into discrete lag by increasing p and a indefinitely while keeping the mean lag $(p+1)/a$ fixed. It should however be noted that this sharpening of the distribution requires quite high p, and that for p below 10 the width of the distribution is of order p/a.

2b Instantaneous Models

The method adopted in this chapter to study stability of differential equations uses equations linearised about the equilibrium point. It is assumed that the stability of these linearised equations ensures the stability, against small disturbances from the equilibrium, of the non-linear equations. Global stability is not examined. This method is standard in the case of ordinary differential equations and delay-differential equations. For integro-differential equations results justifying this method have recently been obtained [Miller (1972), Cushing (1975)].

Let the equations (1) have equilibrium point $\underset{\sim}{x}^0$, at which all the g_i are zero. Define new variables

$$\underset{\sim}{X} = \underset{\sim}{x} - \underset{\sim}{x}^0.$$

These new variables satisfy, to lowest order in X_i, a linear equation

$$\frac{d\underset{\sim}{X}}{dt} = A\underset{\sim}{X}, \tag{9}$$

in which A is the Jacobian matrix,

$$A^{ij} = \frac{\partial g_i}{\partial x_j} . \tag{10}$$

This equation describes the dynamics of small deviations from equilibrium. It has solutions with exponential time dependence,

$$X_i = X_i^o \exp(\lambda t),$$

where X_i^0 are initial displacements, and λ is any root of the polynomial equation

$$P(\lambda) = |A - \lambda I| = 0. \tag{11}$$

The equilibrium point is asymptotically stable against small displacements if all the roots λ have negative real part. It is marginally stable if one root has zero real part and all the rest have negative real part, and unstable if any root has positive real part. Writing the equation (11) in the form

$$\pm P(\lambda) = \lambda^n + a_1\lambda^{n-1} + . . + a_{n-1}\lambda + a_n = 0, \tag{12}$$

the presence of a negative coefficient a_r implies that there is a real positive root. When all the coefficients are positive there are various ways of testing whether all real parts of the roots are negative. The standard method is to evaluate alternate members of the set of Routh-Hurwitz determinants,

$$\Delta_{m+1} = \begin{vmatrix} a_1 & a_3 & a_5 & . & a_{2m+1} \\ 1 & a_2 & a_4 & . & a_{2m} \\ 0 & a_1 & a_3 & . & a_{2m-1} \\ 0 & 1 & a_2 & . & a_{2m-2} \\ . & . & . & . & . \end{vmatrix}$$

If any of these are negative the equilibrium point is locally unstable. Thus for $n = 2$, positive coefficients a_1 and a_2 suffice for stability, for $n = 3$ Δ_2 must be positive,

$$a_1a_2 - a_3 > 0,$$

and for $n = 4$ Δ_3 must be positive,

$$\Delta_2 a_3 - a_1^2 a_4 > 0.$$

Beyond $n = 4$, at least two of these determinants must be evaluated. With increasing n the method soon becomes tedious, especially if one seeks results that display how stability depends on the parameters of the model, rather than working with particular sets of numerical values of these parameters.

It is sometimes possible to obtain explicit results for any n by searching for roots of (11) with zero real part. This will be illustrated in later Chapters. One may consult the book by Porter (1967) for the Routh-Hurwitz method and others.

Some insight into the stability analysis may be gained from the method of loop analysis, introduced into the context of biological models by Levins (1975). One associates with the matrix A a directed signed weighted graph, to be called for short a <u>digraph</u>. Each variable label i defines a <u>vertex</u>. Each non-zero interaction A^{ij} defines an <u>edge</u>, directed from i to j. The weight of this edge is $|A^{ij}|$, and its sign is the sign of A^{ij}. Diagonal terms A^{ii} define <u>loops</u> of length 1, and pairs of reciprocal interactions, A^{ij} and A^{ji}, define loops of length 2. A sequence of q non-zero interaction terms A^{ij}, A^{jk}, . . . A^{ri}, with no two intermediate labels the same, defines a loop of length q. Loops are termed <u>disjunct</u> if they have no vertex in common.

Figure 1 shows three examples of digraphs. I corresponds to the matrix

$$\begin{pmatrix} A^{11} & A^{12} & 0 \\ A^{21} & A^{22} & A^{23} \\ A^{31} & 0 & 0 \end{pmatrix}$$

II to the matrix

$$\begin{pmatrix} A^{11} & A^{12} & 0 & A^{14} \\ A^{21} & 0 & 0 & 0 \\ 0 & 0 & 0 & A^{34} \\ 0 & 0 & A^{43} & 0 \end{pmatrix}$$

and III to the matrix

$$\begin{pmatrix} 0 & A^{12} & A^{13} & A^{14} \\ A^{21} & 0 & 0 & 0 \\ A^{31} & 0 & 0 & 0 \\ A^{41} & 0 & 0 & 0 \end{pmatrix}$$

I has two loops of length 1, one of length 2 and one of length 3. The only disjunct loops are those of length 1. II has 2 disjunct loops of length 2, a single loop of length 1, and an edge, 14, not forming part of a loop. III is an example of a "simple rosette" digraph, with no disjunct loops.

Only edges forming part of one or more loops contribute to the stability equation (11). For example in the case II the isolated element A^{14} does not appear

FIG 1

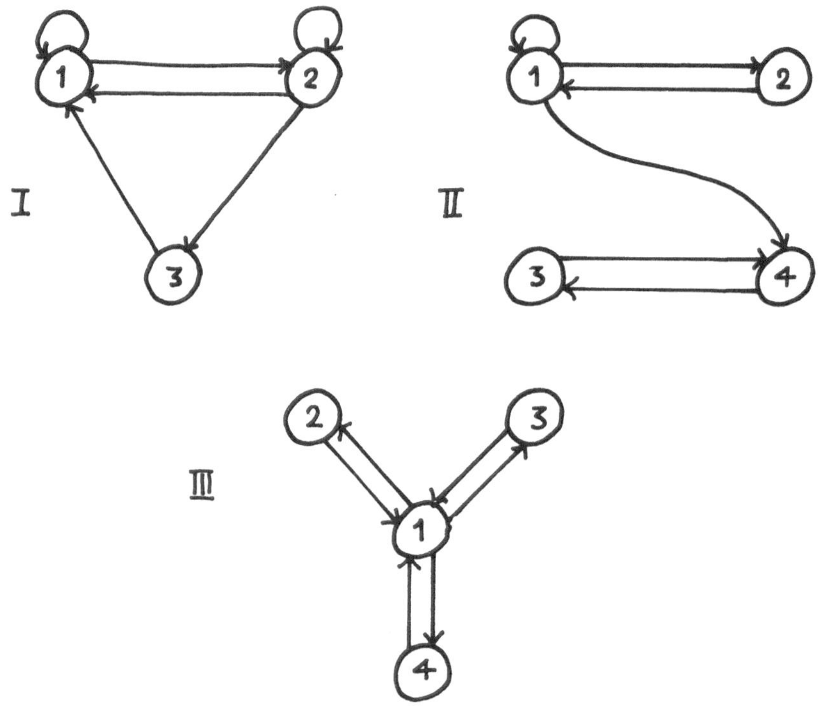

in

$$|A-\lambda I| = [-\lambda(A^{11}-\lambda)-A^{12}A^{21}]\,[\lambda^2-A^{34}A^{43}].$$

Setting down the stability equation is a question of keeping track of the loops. The determinant $|A|$ can be expressed as

$$\sum_{p,q} (-)^{p-q}\, L(p,q), \tag{13}$$

where $L(p,q)$ is the product of p interaction terms making up q disjunct loops. They must be disjunct because no term of the determinant contains a repeated label. The sum here is over all $q \leqslant p$ and over all possible products. Levins defines the "feedback" at level s as

$$F_s = \sum_q (-)^{q+1} L(s,q). \tag{14}$$

For example $F_1 = \text{TrA}$, and

$$F_2 = \sum_{j \neq i} A^{ij} A^{ji} - \sum_{i,j} A^{ii} A^{jj}.$$

He shows that the stability polynomial of equation (12) can be written as

$$\pm P(\lambda) = \lambda^n - \sum_{s=1}^{n} F_s \lambda^{n-s}. \tag{15}$$

Thus there is a precise correspondence between positive feedback and negative, and consequently destabilising, coefficient a_s. The equilibrium point can be unstable either through positive feedback at some level, or through an excess of negative feedbacks at high levels relative to those at low levels, revealed by the sign of the Routh-Hurwitz determinants.

The complications of the stability analysis stem from the multiple counting implied by (14) and from the appearance of disjunct loops. Models with no disjunct loops - rosette digraphs - or only those of length 1, can be relatively simple to analyse. I have discussed the question of lags in such models elsewhere [MacDonald (1977a)]. The context was not a biological one. One can envisage biological models with rosette structure, such as a population model of parasites infesting several final hosts, always through the same intermediate host. However, I have not found a useful biological illustration of lag effects in this way.

From the loop analysis, one can learn something about the effects to be expected when the equations are supplemented by the linear chain. Consider the digraph 1 of Figure 1, to which is to be added a second order lag, that is to say a lag with memory function $G_a^2(u)$. The linear chain adds three new vertices. Figure 2 illustrates some of the ways this can change the digraph. I corresponds to completely replacing x_2 by the delayed form in the equation for $\frac{dx_1}{dt}$. II corresponds to replacing x_2 by the delayed form in one part of the same equation but not in another part. So in I a loop of length 2 is increased to length 5, and one of length 3 is increased to length 6. In II the loop of length 2 remains, and a

FIG 2

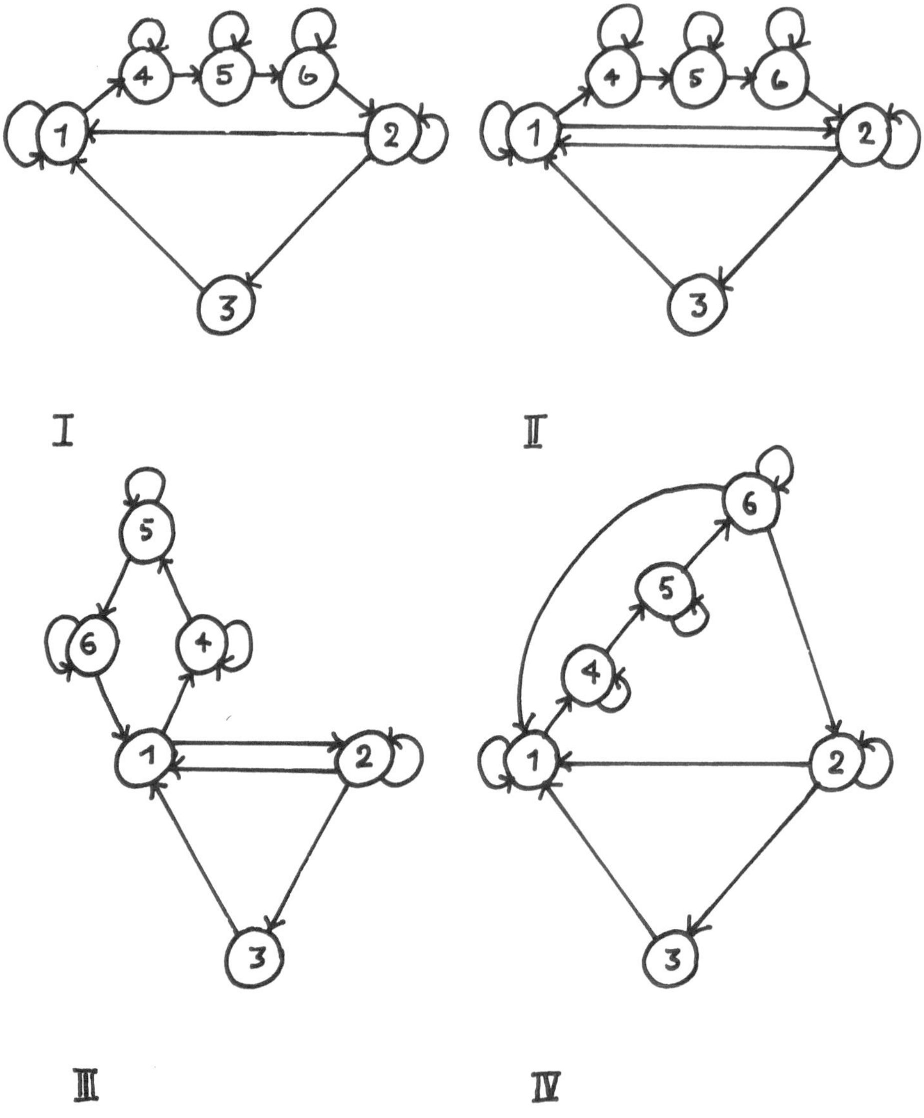

new one of length 5 is added in parallel; similarly the loop of length 3 remains, and one of length 6 is added in parallel. III corresponds to replacing x_1 in the equation for $\frac{dx_1}{dt}$ by the delayed form. The loop of length 1 is replaced by one of length 4. At each new vertex in any of these cases there is a new loop of length 1, negative sign and weight a. The extended loops do not change sign but their weight contains a factor a^{p+1}.

The final diagram, IV, of Figure 2 illustrates a case in which the first equation is

$$\frac{dx_1}{dt} = x_1 - x_1 x_2,$$

and the whole of the product term is delayed. Here the loop linking vertices 1 and 2 is lengthened, and there is also a new loop beginning and ending at 1.

In specific cases in which there are only a few instantaneous equations, it is best to set up the stability equation by direct calculation of the determinant $|A - \lambda I|$ or, if only one variable is delayed, by first setting down the stability equation for a general distributed lag, as described in section d of this chapter.

2c Models with a Single Discrete Lag

Consider modifying the equations (1) by replacing $x_m(t)$ in part of the k'th equation by the delayed form $\bar{x}_m = x_m(t-T)$. The equilibrium points are unchanged. The equations linearised about an equilibrium point $\underset{\sim}{x}^0$ take the form

$$\frac{d\underset{\sim}{X}}{dt} = \underset{\sim}{\bar{A}}\, \underset{\sim}{X}\,, \tag{16}$$

in which the notation $\bar{A}$ means that the term $A^{km}X_m(t)$ is replaced by

$$\alpha^{km}\, X_m(t-T) + \beta^{km}\, X_m(t),$$

where

$$\alpha^{km} = \frac{\partial \hat{g}_k}{\partial \bar{x}_m}, \qquad \beta^{km} = \frac{\partial \hat{g}_k}{\partial x_m}, \tag{17}$$

so that

$$A^{km} = \alpha^{km} + \beta^{km}. \tag{18}$$

[The notation is analogous to that employed in equation (2).] These equations still have solutions with exponential time dependence, and the stability equation can again be set down, in the form

$$Q(\lambda) = H(\lambda) + K(\lambda)\exp(-\lambda T), \tag{19}$$

where the polynomial $P(\lambda)$ of equation (12) is

$$P(\lambda) = H(\lambda) + K(\lambda), \tag{20}$$

and $P(\lambda)$ is changed to $Q(\lambda)$ by replacing A^{km} by

$$\alpha^{km}\exp(-\lambda T) + \beta^{km}. \tag{21}$$

Since I consider only retarded equations, in which the lag appears in a variable and not in a time derivative, $H(\lambda)$ is of higher order than $K(\lambda)$. This ensures [El'sgol'ts and Norkin (1973)] that, as T tends to zero, n of the roots of (19) tend towards the roots of $P(\lambda) = 0$, while the remainder have real part tending to large negative values. So, as the lag is made small, the stability property of the equilibrium point in the instantaneous model is recovered.

One can approach the problem of establishing the sign of the real part of the roots of (19) in various ways. One is to approximate $\exp(-\lambda T)$ by a rational fraction and multiply (19) throughout by the denominator of this fraction. This converts the problem into one with a polynomial equation once more, and the Routh-Hurwitz method can then be used. The Padé approximants give a systematic way of improving this approximation by means of polynomials of increasing order. Writing the (a/b) Padé approximant to $\exp(-\lambda T)$ as

$$N(a, b, \lambda T)/D(a, b, \lambda T),$$

the new polynomial equation is

$$H(\lambda)\, D(a,b,\lambda T) + K(\lambda)\, N(a,b,\lambda T) = 0. \tag{22}$$

The two polynomials defining the Padé approximant are

$$N(a,b,x) = 1 + \sum_{n=1}^{a} (-)^n \frac{x^n}{n!} M(a,b,n)$$

and

$$D(a,b,x) = 1 + \sum_{n=1}^{b} \frac{x^n}{n!} M(b,a,n),$$

with

$$M(a,b,n) = \prod_{m=0}^{n-1} \frac{a-m}{a+b-m}.$$

This method, or a series approximation to $\exp(-\lambda T)$, can at best approximate the roots of (19) with lowest modulus. It is therefore useful only if one can be sure that the first roots to acquire positive part as T increases are in fact those of lowest modulus. An example of this method will be shown, and compared with straightforward series expansion, in Chapter 4. Holst (1969) discusses how to use Padé approximants in this context.

A generally more useful method is to find the conditions for a pair of roots of equation (19) to be pure imaginary at finite T. One examines the equation

$$H(i\omega_o) + K(i\omega_o)\,[\cos(\omega_o T) - i\,\sin(\omega_o T)] = 0. \tag{23}$$

The real and imaginary parts must both be zero so there are two equations from which to determine ω_0 and T. Since the circular functions appear in the conditions derived from equation (23), one may expect stability and instability to alternate as T increases. Generally the main question is whether instability is possible, and the lowest T for which instability occurs is sought. The general theory is given by Pontriagin (1942), and summarised in the books by Bellman and Cooke (1963), Porter (1967) and Hale (1977). A systematic numerical procedure is set out by Hsu (1970).

2d Models with a Single Distributed Lag

Now examine the stability equation obtained when x_m, in the k'th equation of (1), is replaced by

$$\bar{x}_m = \int_{-\infty}^{t} x_m(\tau)\, G(t-\tau)\, d\tau.$$

The linearised equations now take the form

$$\frac{d\underset{\sim}{X}}{dt} = \underset{\sim}{A'\,X}, \tag{24}$$

in which the notation A' means that the term $A^{km}X_m(t)$ is replaced by

$$\alpha^{km}\,\bar{X}_m + \beta^{km}\,X_m, \tag{25}$$

with, as in 2c,

$$\alpha^{km} + \beta^{km} = A^{km}. \tag{26}$$

These linearised equations still have solutions with exponential time dependence, and the stability equation takes the form

$$H(\lambda) + K(\lambda)\,F(\lambda). \tag{27}$$

Here $F(\lambda)$ is the Laplace transform of the memory function,

$$F(\lambda) = \int_0^{\infty} G(u)\,\exp(-\lambda u)\,du. \tag{28}$$

The change from $P(\lambda)$ to $Q(\lambda)$ is now made by replacing A^{km} by

$$\alpha^{km}\,F(\lambda) + \beta^{km}. \tag{29}$$

Once more $H(\lambda)$ is of higher order than $K(\lambda)$. If $G(u) = \delta(u-T)$ one recovers the form of $Q(\lambda)$ of the previous section, since the Laplace transform becomes $\exp(-\lambda T)$.

The stability equation (27) becomes amenable to the techniques of 2b and 2c if $F(\lambda)$ is either a sum of rational fractions in λ, or in $\exp(-\lambda T)$, or in both. I shall make use of the following two forms:

I. Square lag, for which the memory function is

$$G(u) = 1/T, \quad u \leq T,$$

$$G(u) = 0, \quad u > T,$$

and the Laplace transform is

$$F(\lambda) = (1 - \exp(-\lambda T))/\lambda T.$$

The stability equation is

$$H(\lambda)\,\lambda T + K(\lambda) - K(\lambda)\exp(-\lambda T) = 0. \tag{30}$$

II. The linear chain. With the memory function

$$G_a^p(u) = \frac{a^{p+1}u^p}{p!}\exp(-au),$$

the Laplace transform is

$$F(\lambda) = \left(\frac{a}{a+\lambda}\right)^{p+1} \tag{31}$$

the stability equation is

$$H(\lambda) + K(\lambda)\left(\frac{a}{a+\lambda}\right)^{p+1} = 0. \tag{32}$$

So long as one works with finite a the roots of equation (32) are the roots of the polynomial equation

$$H(\lambda)(a+\lambda)^{p+1} + a^{p+1}K(\lambda) = 0. \tag{33}$$

This is the form obtained for the stability equation if the linear chain trick is employed as described in 2a. The factor a^{p+1} comes from the stretched edge effect shown in Figure 2, and the factor $(a+\lambda)^{p+1}$ comes from the set of new stabilising loops of length 1 at each of the new vertices.

Two additional special cases may be mentioned, while others may be found in the paper by Lewis (1977).

III. Linear combinations. Since the Laplace transform is a linear operator, one can use sums of memory functions G_a^p with different values of a and p, simply summing the corresponding $F(\lambda)$.

IV. Memory function with a gap, for example

$$G(u) = 0, \quad u \leq T,$$

$$G(u) = G_a^p(u-T), \quad u > T.$$

For this the Laplace transform is

$$F(\lambda) = \exp(-\lambda T) \left(\frac{a}{a+\lambda}\right)^{p+1} . \tag{34}$$

Two topics need to be examined in connection with the linear chain case. One is the question of the limits of the stability equation as $a \to \infty$ (short mean lag) or as $a \to 0$ (long mean lag). The other is the use of convoluted memory functions.

The limiting form of equation (32) as $a \to 0$ is

$$H(\lambda) = 0.$$

This is the stability equation for the digraph from which the edge containing the lag has been removed. [A connection through which a signal takes an infinitely long time to pass may as well not be there.]

To determine what happens to the roots of (32) as $a \to \infty$, consider first a region of the λ-plane which contains all the roots of (12), and is bounded by the circle $|\lambda| = L$. Then given a small positive number ε one can find $A(L,\varepsilon)$ such that $a > A$ implies that in and on this circle

$$1 - \left|\left(\frac{a}{a+\lambda}\right)^{p+1}\right| < \varepsilon.$$

One can then apply Rouché's theorem to $P(\lambda)$ and

$$R(\lambda) = K(\lambda) \left\{\left(\frac{a}{a+\lambda}\right)^{p+1} - 1\right\},$$

to conclude that (32) has n roots within $|\lambda| = L$. By similar reasoning one can show that there is a root of (32) as near as one wishes to each root of (12).

Now consider roots of (32) that are assumed to increase in modulus as a increases. I suspect that they must have negative real part, and do not affect the stability analysis in the limit of $a \to \infty$. It would be rather anomalous if in this respect alone distributed lag were to be more destabilising than discrete lag. An attempt at a proof of this can be made, assuming that one can write any of these roots as a dominant term $a^s e^{i\phi}$, where $s > 0$, plus terms with modulus less strongly increasing with a. Then (32) gives

$$\lambda = -a + a \left[\frac{K(\lambda)}{H(\lambda)}\right]^{1/(p+1)},$$

and taking m from 0 to p this gives approximately

$$\lambda_m = -a + \alpha a^{\left(1 - \frac{S(N-M)}{p+1}\right)} e^{\frac{mi\ \phi(M-N)}{p+1}} \left[\frac{1 + \beta_1\lambda^{-1} + }{1 + \gamma_1\lambda^{-1} + }\right]^{1/p+1}.$$

where $H(\lambda)$ is taken to be $\lambda^N + \beta_1\lambda^{N-1} + \ldots + \beta_N$ and $K(\lambda)$ to be $\alpha^{p+1}\ (\lambda^M + \gamma_1\lambda^{M-1} + \ldots + \gamma_M)$. The factor in square brackets can be made as close as one wishes to 1 by raising a. Hence the roots approach

$$\lambda_m = -a + \alpha b,$$

where

$$|b| = a^{\left[1 - \frac{S(N-M)}{p+1}\right]},$$

and $|b|$ is of lower order than 1 in a, since $s > 0$ and $N > M$. So the lowest order approximation to these p+1 roots is $\lambda_m = -a$, and the next lowest approximation is

$$\lambda_m = -a + \alpha a^{\frac{M-N+1}{p+1}} \exp\left(\frac{im\phi(M-N)}{p+1}\right). \qquad (35)$$

As mentioned in 1b it may be necessary to use convoluted lags, so that the delayed variable is

$$z = \int_{-\infty}^{t} G_2(t-\tau_2)d\tau_2 \int_{-\infty}^{\tau_2} G(\tau_2-\tau_1)\ x(\tau_1)d\tau_1. \qquad (36)$$

Now the Laplace transform of the convolution of two memory functions is the product of the transforms of the two functions. Thus if

$$G_1 = G_a^p, \quad G_2 = G_a^q,$$

the transform to be used in the stability equation is

$$F(\lambda) = \left(\frac{a}{a+\lambda}\right)^{p+q+2}.$$

This is of course the transform of G_a^{p+q+1}. So the delayed variable is the same as one would obtain by using a single lag with order p+q+1 and the same parameter a as in the convoluted lags. This can also be seen directly by evaluating the inner integral in (36).

When the two lags are

$$G_1 = G_a^p, \quad G_2 = G_b^q,$$

the equivalent single lag is not so simply related to these functions, but the stability equation can again be readily written, with the transform

$$F(\lambda) = \frac{a^{p+1}\, b^{q+1}}{(a+\lambda)^{p+1}\,(b+\lambda)^{q+1}} .$$

Convolutions of a rather similar type are used in queuing theory. There the convention used is to keep the mean lag constant as one convolutes successive lags, resulting in the use of the Erlang function, which in the notation used here is G_p^{p-1}. In the present context it is more natural to add the mean lags as successive lags are convoluted.

2e <u>An Inequality for Distributed Lag</u>

In this section I return to equation (27) with general $F(\lambda)$. To find whether this equation can have a pure imaginary pair of roots, one has to seek a solution of

$$L(i\omega_o) = \frac{H(i\omega_o)}{K(i\omega_o)} = -F(i\omega_o). \qquad (37)$$

A necessary condition for this is

$$|L(i\omega_o)| \leqslant 1.$$

For, by the definition of the transform,

$$F(i\omega_o) = \int_0^\infty G(u)\, \exp(-i\omega_o u)du,$$

and this implies

$$|F(i\omega_o)| \leqslant \int_0^\infty G(u)\, |\exp(-i\omega_o u)|du,$$

when G(u) is taken to be positive for all u. This last expression is equal to

$$\int_0^\infty G(u)du,$$

which is equal to one, by the normalisation adopted for the memory function. So it may be possible, in a particular model, to show that

$$|L(i\omega_o)| > 1,$$

and thus exclude the possibility of a destabilising lag. This method will be applied in section g of this chapter, and in Chapter 5.

The extreme case of discrete lag gives

$$|F(i\omega_o)| = 1.$$

So if a discrete lag cannot destabilise an equilibrium point in the model, no distributed lag can do so either.

2f The Monod Chemostat Model

To illustrate the stability analysis I shall examine two models in which the instantaneous equations involve two variables. One is the Monod model of the growth of the population of a simple organism in a chemostat. The other is a particular version of a model of May (1976) for obligate mutualism between a plant species and an insect species. Other examples employing the equation (32) with low p values can be found in another paper by May (1973). May stresses that these distributed lags have advantages both of realism and of convenience as compared with discrete lags.

The chemostat is a device for maintaining a constant population of a simple organism by maintaining a constant flow of dissolved nutrient through a vessel. This has advantages, as compared with the use of a closed vessel, for studying behaviour to be expected in such natural environments as estuaries or the human mouth. Also in industrial microbiology a continuous flow process is usually to be preferred to a batch process. [Dean et al. (1976)]

In the Monod model [Monod (1950)] the two variables are the population x(t) of the organism and the amount y(t) of nutrient. A steady flow of water through the vessel allows the addition of nutrient at a constant rate $K\bar{y}$, and washes out both nutrient and organisms at rates proportionate to the amounts present, -Kx and -Ky. The depletion of the nutrient due to uptake by the organisms, and the consequent growth of the population, are taken to be proportional to a non-linear function of the form

$$\frac{xy}{\gamma + y} .$$

When there is little nutrient present the organisms each take up nutrient at a rate proportional to the amount present. When there is abundance of nutrient the uptake by each organism saturates. The instantaneous equations of the Monod model are therefore

$$\frac{dx}{dt} = -Kx + \frac{\mu xy}{\gamma+y}, \tag{38}$$

$$\frac{dy}{dt} = K(\bar{y}-y) - \frac{\mu' xy}{\gamma+y}. \tag{39}$$

The equilibrium points are

$$x = 0, \quad y = \bar{y},$$

which is of no further concern here, and

$$x = x^o = (\mu/\mu')(\bar{y}-y),$$

$$y = y^o = K\gamma(K-\mu).$$

The Jacobi matrix A defined relative to this second equilibrium point has elements

$$A^{11} = 0. \quad A^{12} = \frac{\mu x^o \gamma}{(\gamma+y^o)^2} \overset{\text{def}}{=} \frac{\mu}{\mu'}\alpha.$$

$$A^{21} = \frac{-\mu' y^o}{\gamma+y^o} \overset{\text{def}}{=} -\frac{\mu'}{\mu}\beta. \quad A^{22} = -\alpha-\beta.$$

Hence the stability equation (12) is

$$\lambda^2 + \lambda(\alpha+\beta) + \alpha\beta = 0. \tag{40}$$

This equilibrium point is always stable.

If x(t) is interpreted as the biomass of the organism there can be no lag, because of energy conservation. However if x(t) is interpreted as the number of organisms, then lag may be introduced as a crude way of allowing for the relation between the growth of individual single cell organisms and their likelihood of splitting. Caperon (1969) and Thingstad and Langeland (1974) discuss the dynamics of the chemostat in terms of number of organisms, and examine the effects of lag.

The lag occurs in the non-linear term in equation (38). I take the lag to affect each of the y's appearing in that term, so that in the modified version of equation (40) A^{12} is multiplied by $F(\lambda)$. I first examine discrete lag, using equation (23), which becomes

$$i\omega_o(i\omega_o + \alpha+\beta) + \alpha\beta(\cos(\omega_o T) - i\sin(\omega_o T)) = 0.$$

This equation implies

$$\omega_o^2 = \alpha\beta \cos(\omega_o T), \tag{41}$$

$$\omega_o(\alpha+\beta) = \alpha\beta \sin(\omega_o T),$$

and these equations are compatible for a pair of positive ω_o, T, as can be seen by evaluating $\cos^2(\omega_o T) + \sin^2(\omega_o T) = 1$.

Next I consider square lag, using equation (30) with a pure imaginary root. This gives

$$-\omega_o^2 T(i\omega_o+\alpha+\beta) + \alpha\beta - \alpha\beta \cos(\omega_o T) + i\alpha\beta \sin(\omega_o T) = 0,$$

which implies

$$\begin{aligned} \omega_o^2 T(\alpha+\beta) &= \alpha\beta(1-\cos(\omega_o T)), \\ \omega_o^3 T &= \alpha\beta \sin(\omega_o T). \end{aligned} \tag{42}$$

These are again consistent, for positive ω_o and T, by the same test. Caperon (1969) in calculations in which the parameters of the Monod model were fixed by the equilibrium concentrations observed in his experiments, and for which he made what he took to be a reasonable estimate of the mean lag, found the equilibrium point to be stable for square lag, but unstable for discrete lag. Thingstad and Langeland (1974) also present cases in which the equilibrium point is unstable for discrete lag.

The p = 0 and p = 1 cases of the linear chain trick are readily tested, applying the Routh-Hurwitz method to equation (33). For p = 0 this equation is

$$\lambda(\lambda+\alpha+\beta)(\lambda+a) + a\alpha\beta = 0,$$

and

$$\Delta_2 = (\alpha+\beta+a)(\alpha+\beta)a - \alpha\beta a > 0. \tag{43}$$

So there can be no instability. For p = 1 the equation is

$$\lambda(\lambda+\alpha+\beta)(\lambda+a)^2 + a^2\alpha\beta = 0,$$

and

$$\begin{aligned} \Delta_3 = {} & a^2(\alpha+\beta)\left[(2a+\alpha+\beta)(a^2+2a(\alpha+\beta)) - a^2(\alpha+\beta)\right] \\ & - (2a+\alpha+\beta)^2 \alpha\beta a^2. \end{aligned} \tag{44}$$

The change in sign between the coefficients of a^5 and a^2 indicates that for small enough a (long enough mean lag) the equilibrium point becomes unstable. The change from stable to unstable comes by way of a pure imaginary pair of roots. As will be discussed in the next Chapter, this indicates a possibility that there is a periodic solution of the Monod equations with this form of lag. In fact numerical integration of the equations, using the linear chain, reveals a periodic solution.

Sustained oscillations of a yeast population in a chemostat were observed by Finn and Wilson (1953) and discussed in terms of discrete lag using a linear equation. Since then, there has been little experimental evidence for sustained population oscillations in chemostats. However, in the thesis of A. Cunningham (1977) a lag is used in a generalised Monod model to account for aspects of the transient oscillations of an algal population, obtained by changing the flow rate in a chemostat. It seems likely that a simple lag effect is too crude a way of allowing for the relation between cell size and cell splitting.

One more point about the method can be illustrated in this example. If for some reason it seems more appropriate to use delayed y only in the numerator of the non-linear term in (38), then one writes the Jacobian matrix element out in full as

$$A^{12} = \frac{\mu x^o}{\gamma + y^o} - \frac{\mu y^o x^o}{(\gamma + y^o)^2},$$

and amends only the first part to include the Laplace transform $F(\lambda)$. Thus

$$A'^{12} = \frac{\mu x^o}{\gamma + y^o} F(\lambda) - \frac{\mu y^o x^o}{(\gamma + y^o)^2}$$

and the less tidy stability equation is

$$\lambda^2 + \lambda(\alpha + \beta) - \left(\frac{\mu'}{\mu}\beta\right)\left(\beta\frac{y^o}{x^o}\right) F(\lambda) + \alpha\beta\frac{y^o}{\gamma} = 0.$$

2g May's Model of Obligate Mutualism

Mutualism, as discussed for example in the book by Gilbert and Raven (1975), has received less theoretical attention than other modes of interaction between two species, such as predation, parasitism or competition. This is particularly true of obligate mutualism, in which the presence of each species is not merely beneficial but essential to the other. Early work by Kostitzin (1934), as reviewed by Rescigno and Richardson (1973), refers to a very diluted form of mutualism, in which some members of two competing species opt out of competition. Rescigno and

Richardson also briefly discuss a model in which two species have populations that rise to limiting values, these limiting values being raised in the presence of the other species. Such a model is also discussed by May (1976a) in a more explicit form. May also describes a model of obligate mutualism, and discusses why this has been rather a neglected topic. He points out that quantitative data are sparse and that this comes in part from the fact that obligate mutualism is confined to the tropics and sub-tropics. He also points out that a model of mutualism which is a simple analogue of the Lotka-Volterra models of competition or predation leads to a ridiculous result, that populations either fall to zero or rise without limit.

May offers a model which avoids this dilemma, and which he suggests is relevant to the prevalence of obligate mutualism in the tropics. The model has stable equilibrium points at zero populations and at a pair of finite population values. At a somewhat lower pair of population values there is an unstable equilibrium point. Moderate displacements from the upper equilibrium are corrected, but larger displacements, which put the population values into the region of attraction of the origin, can not be corrected. Accepting the common assumption that typical tropical environments are more stable and more predictable than those of temperate regions, the model suggests that mutually dependent pairs of species are too vulnerable to survive in temperate regions. May gives a qualitative description of his model, and that given here is a specific example consistent with his description.

Both May (1976a) and Gilbert and Raven (1976) stress the possible importance of time lag in mutualism. The most common type of mutualism is that between a plant species and a pollinating insect species, which have co-evolved until each is uniquely suited to the other. Successful pollination does not increase the plant population until the seeds have had time to be deposited and to sprout. The nectar gathered by the insect may be used to feed larvae, leading to the appearance of the next generation of insects, again with a delay. One may reasonably suppose that it is typical for the two lags to be on average the same, an annual plant being associated with an insect having one generation each year.

A possible simple model of obligate mutualism has as variables the two populations X and Y, each constrained to lie between 0 and some upper limit, which is proportional to the excess of the other population over some threshold value. If either population takes a value below its threshold, the other falls to 0. This can be expressed, for example, by the equations

$$\frac{dX}{dt} = \delta X \left[Y-\beta-\gamma X\right], \tag{45a}$$

$$\frac{dY}{dt} = \varepsilon Y \left[X-\alpha-\phi Y\right]. \tag{45b}$$

The form below threshold is rather arbitrary. Equations (45) have a smooth transition across the thresholds and save writing alternative forms above and below the thresholds. The origin (0,0) is a stable equilibrium point. The other equilibrium point, given by

$$X_o - \phi Y_o = \alpha,$$

$$Y_o - \gamma X_o = \beta,$$

lies above the threshold values of X and Y and is unstable. In this model low populations die out and high ones rise indefinitely. May's model improves on this simple one by providing a second stable equilibrium point to trap the rising trajectories. This is done by replacing one or the other of the linear factors $Y-\beta-\gamma X$, $X-\alpha-\phi Y$, by a non-linear factor. Choosing the first of these, the general nature of the model can be shown by plotting the curve $\frac{dX}{dt} = 0$ and the line $\frac{dY}{dt} = 0$ as in Figure 3.

I shall examine a particular version of this model, which keeps (45b) unchanged, and replaces (45a) by

$$\frac{dX}{dt} = \delta X \left[\sqrt{Y-\beta} - \gamma X\right], \quad Y > \beta, \tag{45c}$$

$$\frac{dX}{dt} = -\delta X \left[\sqrt{\beta-Y} + \gamma X\right], \quad Y < \beta. \tag{45d}$$

Again the form below threshold is chosen to fit smoothly; one might just as well take a pure exponential decay. The equilibrium points are now (0,0), (X^+,Y^+) and (X^-,Y^-), where

$$X^{\pm} = \frac{1}{2\gamma^2\phi}\left[1 \pm \surd\{1 - 4(\alpha+\beta\phi)\,\phi\gamma^2\}\right],$$
$$Y^{\pm} = (X^{\pm}-\alpha)/\phi. \tag{46}$$

To ensure that X^+, X^- are real the parameters must satisfy

$$4(\alpha+\beta\phi)\phi\gamma^2 < 1.$$

It is readily verified that X^+, X^- are greater than α and Y^+, Y^- are greater than β.

Lags can enter diagonally, for example the rate of emergence of new adult insects depending on the size of the adult insect population of the previous generation, or off-diagonally, this rate depending on the size of the plant population available to the previous insect generation. It is also clear that it is

FIG 3

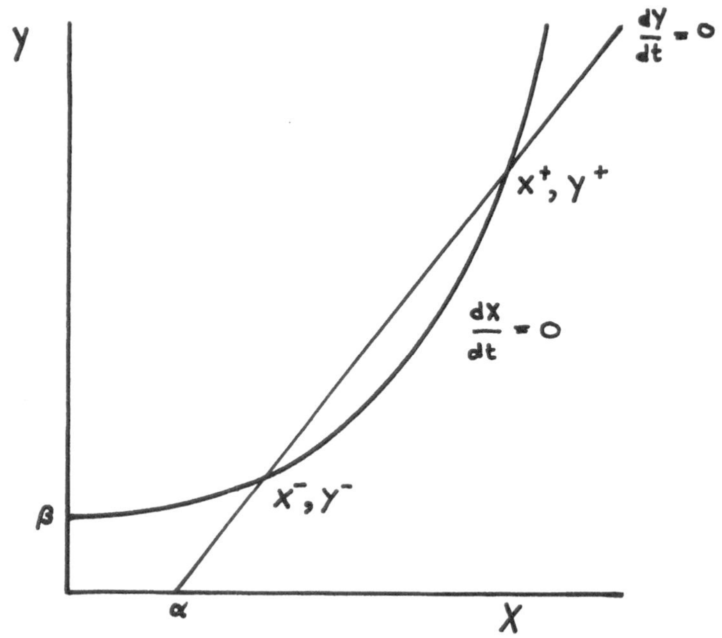

necessary to divide $\frac{dX}{dt}$ and $\frac{dY}{dt}$ up into birth and death terms before introducing the lags. In this context there seems to be no reason to expect a lag in the death terms. This is taken also to imply that lag is not inserted in the equations (45) below threshold. The simplest way to divide up the equations is to take the parts in X^2 and Y^2 as death terms, and the other parts as birth terms. So the most useful equations to study are the pair

$$\frac{dX}{dt} = \delta\bar{X}\,(\sqrt{\bar{Y}-\beta}) - \delta\gamma X^2,$$

$$\frac{dY}{dt} = \epsilon\bar{Y}\,(\bar{X}-\alpha) - \epsilon\phi Y^2, \tag{47}$$

where the bar indicates a delayed variable.

In order to illustrate the methods of stability analysis I shall look first at two simpler cases, namely diagonal lag in the equation for $\frac{dX}{dt}$ only, and off-diagonal lag in this equation only. It is helpful first to simplify the notation by scaling the variables X and Y and the time t, by using

$$x = X/\alpha, \quad y = Y/\beta, \quad \epsilon\alpha = 1.$$

No significant generality is lost by setting $\delta\sqrt{\beta} = 1$, so the instantaneous equations I shall use in the remainder of this section are

$$\frac{dx}{dt} = x\left[\sqrt{y-1} - gx\right], \tag{48}$$

$$\frac{dy}{dt} = y\left[x-1-fy\right], \tag{49}$$

in which $g = \gamma\alpha/\sqrt{\beta}$ and $f = \phi\beta/\alpha$. The equilibrium points are given by

$$x^{\pm} = \frac{1}{2fg^2}\left[1 \pm \sqrt{\{1 - 4(1+f)\,fg^2\}}\right],$$

$$y^{\pm} = (x^{\pm}-1)/f.$$

The equation determining local stability is

$$\begin{vmatrix} -gx^{\pm}-\lambda & 1/2g \\ y^{\pm} & -fy^{\pm}-\lambda \end{vmatrix} = 0. \tag{50}$$

The constant term here is

$$y^{\pm}(fgx^{\pm}-1/2g),$$

implying stability for (x^+,y^+) but instability for (x^-,y^-).

The modified form of the first equation with diagonal lag is

$$\frac{dx}{dt} = z\,\sqrt{y-1} - gx^2, \tag{51}$$

and that with off-diagonal lag is

$$\frac{dx}{dt} = x\,(\sqrt{z-1} - gx). \tag{52}$$

In the first case the stability equation is

$$(\lambda+fy^{\pm})(\lambda+2gx^{\pm}-gx^{\pm}F(\lambda)) - \frac{y^{\pm}}{2g} = 0, \tag{53}$$

and in the second case it is

$$(\lambda+fy^{\pm})(\lambda+gx^{\pm}) - y^{\pm}F(\lambda)/2g = 0. \tag{54}$$

The constant term in each of these equations, when distributed lag is taken with

$$F(\lambda) = \left(\frac{a}{a+\lambda}\right)^{p+1},$$

and the equation multiplied through by $(a+\lambda)^{p+1}$, is the same as that in equation (50), except for a factor a^{p+1}. So for (x^-,y^-) this term is always negative, and this equilibrium point is unstable however long the mean lag.

Now consider off-diagonal lag in the case of (x^+, y^+). It is clear that equation (54) can not have a zero root. For it to have a pure imaginary root $+i\omega_o$ the condition is

$$(i\omega_o+fy^+)(i\omega_o+gx^+) = F(i\omega_o)y^+/2g.$$

Now

$$|(i\omega_o+fy^+)(i\omega_o+gx^+)| \geqslant fgx^+y^+ > y^+/2g,$$

and, as discussed in 2e,

$$|F(i\omega_o)| \leqslant 1.$$

So the higher equilibrium point can not be destabilised by an off-diagonal lag.

Things are not so clear cut in the case of diagonal lag and (x^+,y^+). Testing for stability, whether with discrete lag, $F(\lambda) = e^{-\lambda T}$, or distributed lag with $F(\lambda) = (a/(a+\lambda))^{p+1}$ and low values of p, becomes rather cumbersome, and is hardly worthwhile unless one has some idea of acceptable f, g and a or T in a particular context. This is even more the case for the equations (47), which combine diagonal and off-diagonal lag. I shall however examine how the stability equation is set up in this case, since this will bring out another aspect of the analysis of lag equations. The simplified form of the equations (47) is

$$\begin{aligned}\frac{dx}{dt} &= z\sqrt{(w-1)} - gx^2,\\ \frac{dy}{dt} &= w(z-1) - fy^2,\end{aligned} \tag{55}$$

where z and w are the delayed forms of x and y respectively. In the simplest possible treatment (p = 0) these satisfy two additional equations,

$$\frac{dz}{dt} = a(x-z),$$

$$\frac{dw}{dt} = b(y-w).$$

The stability determinantal equation is therefore

$$\begin{vmatrix} -\lambda-2gx^+ & 0 & gx^+ & 1/2g \\ 0 & -\lambda-2fy^+ & y^+ & fy^+ \\ a & 0 & -a-\lambda & 0 \\ 0 & b & 0 & -b-\lambda \end{vmatrix} = 0. \tag{56}$$

In equations (55) it is clear that the average over all earlier times is being performed separately for the two populations. However the coincidence of plants and pollinators at earlier times is necessary for propagation of the new generation. So a more appropriate pair of equations are

$$\frac{dx}{dt} = u-gx^2,$$
$$\frac{dy}{dt} = v-fy^2, \tag{57}$$

in which the nonlinearity in the interaction terms has been shifted into the equations satisfied by the new variables, u and v. The variable u is defined by the integral

$$u = \int_{-\infty}^{t} x(T)\ \sqrt{(y(T)-1)}\ G_a^p(t-T)\,dT,$$

and it must satisfy, for p = 0, the equation

$$\frac{du}{dt} = a(x\ \sqrt{(y-1)}-u).$$

In the same way, the variable v is defined by the integral

$$v = \int_{-\infty}^{t} y(T)\ (x(T)-1)\ G_b^p(t-T)\,dT,$$

and it must satisfy, for p = 0, the equation

$$\frac{dv}{dt} = b(y(x-1)-v).$$

The determinantal stability equation is now

$$\begin{vmatrix} -\lambda-2gx^{+} & 0 & 1 & 0 \\ 0 & -\lambda-2fy^{+} & 0 & 1 \\ agx^{+} & a/2g & -a-\lambda & 0 \\ by^{+} & bfy^{+} & 0 & -b-\lambda \end{vmatrix} = 0 \qquad (58)$$

which turns out to be identical with equation (56).

Since a value of $p \geqslant 1$ would be needed to describe the kind of lag envisaged in this context, exploration of stability, even with $a = b$, is too complicated to be worth entering on unless one can make reasonable choices for the parameters for a particular example. If data were adequate to do this, one might also be in a position to choose between the equations (45) and some other form sharing the general properties implied by Figure 3.

PERIODIC SOLUTIONS

3a Periodic Solutions of the Linear Chain Equations

In the following chapters I shall largely be concerned with the question of periodic behaviour brought about by lags. For distributed lags I shall rely on results obtained, either analytically or numerically, on periodic solutions of the linear chain equations. It therefore becomes necessary to look again at the relationship between the solutions of these equations and the solutions of the related integro-differential equation. So far as analytical results are concerned, and for the bulk of the numerical results also, it suffices to look at a single integro-differential equation. In the notation of Chapter 1,

$$\frac{dx}{dt} = f(x,z), \tag{1}$$

where z is defined by

$$z = \int_{-\infty}^{t} x(\tau)\, G(t-\tau)\, d\tau, \tag{2}$$

and the memory function G(u) is defined by

$$G_a^p(u) = \frac{a^{p+1} u^p}{p!} \exp(-au). \tag{3}$$

The linear chain equations consist of (1) and the equations

$$\begin{aligned} \frac{dz}{dt} &= a(y-z), \\ &\cdots\cdots\cdots\cdots \\ \frac{dv}{dt} &= a(x-v), \end{aligned} \tag{4}$$

or

$$\left\{\frac{d}{dt} + a\right\}^{p+1} z = ax. \tag{5}$$

Let us assume that a periodic solution of (1), (5) has been found, denoted by (z(t), x(t)). Then z(t) is a periodic solution of the linear p+1 order differential equation (5) with periodic forcing term x(t). Now for given periodic forcing term x(t) this linear equation has a unique periodic solution given by the form (2) for z(t), with the kernel in the integral defined by (3). It follows that x(t) is a periodic solution of the integro-differential equation.

The linear chain trick makes available methods from the theory of periodic solutions of sets of ordinary differential equations for applications in delay problems. Since the method always increases the dimension of the set of first order equations, methods special to two dimensions need not be considered. When suitable methods for n dimensions can be identified, the fact that p+1 of the equations are linear tends to facilitate their use. I confine my attention to methods that have been presented as explicit algorithms. For additional potentially useful methods see the review by Cronin (1977). The book by Cushing (1978) gives some methods directly applicable to integro-differential equations.

I shall describe two methods, which will be applied to specific problems in Chapters 4 and 5. The method of Hastings, Tyson and Webster gives conditions for instability of an equilibrium point to imply the existence of at least one periodic solution. The method of Hopf bifurcation gives results concerning the existence of such a solution for parameter values near those corresponding to the transition from stability to instability.

3b <u>The Method of Hastings, Tyson and Webster</u>

Hastings et al. (1977) were in the first instance concerned with a particular problem, that of the Goodwin oscillator model, for which at the time of their work there was evidence of periodic solutions from numerical work. The form of the model is a single non-linear equation

$$\frac{dx_1}{dt} = f(x_n) - b_1x_1, \tag{6}$$

and a chain of linear equations,

$$\frac{dx_r}{dt} = b_rx_r + k_rx_{r-1}, \quad r = 2, .. n. \tag{7}$$

They generalise this to allow the 2cd, . . n'th equations to have non-linearities, so long as these equations reduce to the form (7) on linearising around the equilibrium point.

The kind of argument that these authors employed can best be illustrated by the three-dimensional version of (6, 7) examined earlier by Tyson (1975). In his notation the equations are

$$\frac{dX}{dt} = (1+Z^m)^{-1} - \alpha X,$$

$$\frac{dY}{dt} = X - \beta Y,$$

$$\frac{dZ}{dt} = Y - \gamma Z,$$

or

$$\frac{d^3Z}{dt^3} + (\alpha+\beta+\gamma)\frac{d^2Z}{dt^2} + (\alpha\beta+\beta\gamma+\gamma\alpha)\frac{dZ}{dt} + \alpha\beta\gamma Z = (1+Z^m)^{-1}. \tag{8}$$

The equilibrium point is

$$X^o = Y^o = \beta\gamma Z^o,$$

with Z^o the unique positive root of

$$\alpha\beta\gamma(Z^m + 1)Z = 1.$$

Assume that this is an unstable equilibrium point, as is known can be the case for $m \geq 8$. It had earlier been shown [Griffiths (1968)] that all trajectories are inward on the faces of a box with sides parallel to the X,Y,Z axes and with opposite vertices at

$$(0,0,0), \quad (\alpha^{-1}, (\alpha\beta)^{-1}, (\alpha\beta\gamma)^{-1}).$$

This box contains the equilibrium point. The stability equation is

$$\lambda^3+(\alpha+\beta+\gamma)\lambda^2 +(\alpha\beta+\beta\gamma+\gamma\alpha)\lambda + \alpha\beta\gamma + f = 0,$$

with f defined by

$$f = \alpha\beta\gamma m(1-\alpha\beta\gamma Z^o) > 0.$$

When the equilibrium point is unstable the roots of this equation are a real negative one and a complex pair with positive part. So a line can be defined, through the equilibrium point, such that all trajectories that converge towards the equilibrium point approach this line.

Tyson next considers the box as divided into eight smaller boxes with faces parallel to the faces of the large, and sharing a vertex at the equilibrium point. Then the line in question lies in two diagonally opposite small boxes, which are labelled B_7 and B_8. Tyson examines trajectories that cross internal faces of the small boxes, and shows that any trajectory that is at any time in one of the boxes B_1 to B_6 can never get into box B_7 or box B_8, and so can not be drawn in to the equilibrium point. So there are trajectories that can neither escape from the large box nor go to the equilibrium point. Next he shows that any trajectory that at a given time leaves B_1 for B_2 must proceed from box to box in a specific sequence, and must finally reach the common face of B_1 and B_2 once more. Then a fixed point theorem applies, to the effect that among all such trajectories there is

FIG 4

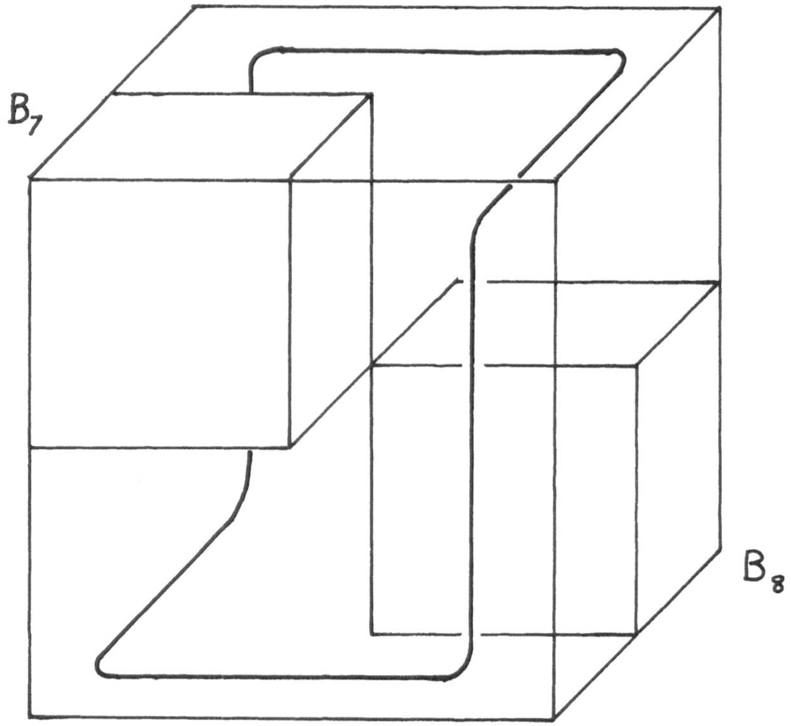

at least one that returns to the same point on the common face of B_1 and B_2 from which it started. This corresponds to a non-constant periodic solution of the equations (8). This situation is illustrated in Figure 4.

As the number of equations in the chain (7) increases, the value of m needed for instability falls, tending to 1 as n becomes very large. If the model is to be applied to sequences of biochemical reactions in cells, it is quite important to be able to deal with large n, and it is significant that implausibly large values of m are not required.

Hastings et al. obtain the generalisation of Tyson's result to n dimensions. Their result can be expressed in terms of a number of tests to be applied to the set of differential equations, without in each case having to find the box and go through the procedure outlined above.

Let there be a set of ordinary differential equations

$$\frac{dx_i}{dt} = g_i(x_1, \ldots x_n), \tag{9}$$

for variables $x_1, \ldots x_n$, defined for $x_i \geq 0$. The restriction to non-negative values is a natural one if the variables are population or cell numbers or chemical concentrations, or the averages over time of such quantities with positive memory functions. Let this set of equations have a unique equilibrium point for positive values of the variables. Let the Jacobian matrix A for these equations, evaluated at this equilibrium point, be of the form

$$\begin{bmatrix} -K_1 & 0 & & & 0 & -d_n \\ d_1 & -K_2 & 0 & & & \\ 0 & d_2 & -K_3 & 0 & & \\ & & & & & \\ 0 & & & & d_{n-1} & -K_n \end{bmatrix} \tag{10}$$

This corresponds to a single closed loop of equations each linking two members of a cyclic sequence of the x_i, as in cases considered in the following chapters in which some of these equations arise from the use of the linear chain trick. Let A have no repeated eigenvalues, that is to say, let the roots of the stability equation be all distinct. Let the following conditions hold for $x_i > 0$:

$$\frac{\partial g_j}{\partial x_j} < 0, \quad \frac{\partial g_j}{\partial x_{j-1}} > 0, \quad 2 \leq j \leq n, \tag{11a}$$

$$\frac{\partial g_1}{\partial x_n} < 0. \tag{11b}$$

Finally, let the following conditions also hold:

$$g_i(0,0) \geq 0, \qquad 1 \leq i \leq n; \tag{12a}$$

$$g_1(x_n,0) \geq 0, \qquad x_n \geq 0; \tag{12b}$$

$$g_1(x_n,x_1) < 0, \qquad x_n > x_n^o, \quad x_1 > x_1^o; \tag{12c}$$

$$g_1(k_n,x_1) > 0, \qquad x_n < x_n^o, \quad x_1 < x_1^o; \tag{12d}$$

$$\frac{\partial g_1}{\partial x_1} \leq C, \qquad x_1 \geq 0, \quad C > 0. \tag{12e}$$

Then if A has any eigenvalues with positive real part there is a positive invariant set,

$$V = \bigcup_i V_i,$$

in the space of the variables x_i, with the V_i arranged in a circuit around the equilibrium point, and such that any trajectory starting in V stays bounded and passes successively and interminably through the V_i, without approaching the equilibrium point, as t increases. Also at least one of these oscillatory trajectories is periodic. The theorem does not state that the oscillatory trajectories necessarily tend to such a periodic trajectory.

It is perhaps worth noting that even quite a simple model incorporating the linear chain, and suspected from numerical integration to have a stable periodic trajectory, or known from the Hopf method to have one in some restricted part of the range of parameters, need not satisfy the conditions of the H.T.W. theorem. An example is the Volterra model of predation with the lag in the non-linear term in the predator rate equation,

$$\frac{dx}{dt} = rx(1-x/K)-\alpha xy,$$

$$\frac{dy}{dt} = -\gamma y+\beta yz,$$

$$\frac{dz}{dt} = a(w-z),$$

$$\cdots\cdots\cdots$$

$$\frac{dr}{dt} = a(x-r).$$

As I shall discuss in Chapter 7 this has a periodic solution, although condition (11a) does not hold. Thus there is scope for weakening the conditions of the theorem of Hastings et al.

3c Hopf Bifurcation

Consider a set of ordinary differential equations, with an equilibrium point at which the roots of the stability equation all have negative part except for a complex pair. For these the sign of the real part changes at a critical value α_c of one of the parameters, α. So long as one looks near enough to the equilibrium point, the trajectories are inward spirals for $\alpha < \alpha_c$, and outward spirals for $\alpha > \alpha_c$. The content of the Hopf bifurcation theorem is that there is always a closed trajectory around the equilibrium point, if one takes α to be near enough to α_c. There are two distinct possibilities, called sub-critical and super-critical Hopf bifurcation.

Sub-critical Hopf bifurcation means that there is a closed unstable trajectory from which trajectories spiral in towards the stable equilibrium point. This can be illustrated by A in Figure 5 and by A in Figure 6. In Figure 5 the amplitude of a periodic solution (an appropriate average distance from the equilibrium point to the closed trajectory) is plotted against increasing α. Solid lines indicate stable regions, either for the equilibrium point or for the periodic solution. The amplitude of the unstable branch of the periodic solution behaves like $(\alpha_c-\alpha)^{\frac{1}{2}}$ near α_c. Figure 6 shows the projection onto a plane of the periodic trajectories and of a typical spiral trajectory, for $\alpha < \alpha_c$ and for $\alpha > \alpha_c$.

Super-critical Hopf bifurcation means that there is a closed stable trajectory with spiral trajectories leaving the unstable equilibrium point and approaching this closed trajectory. This is illustrated by B in Figure 5 and by D in Figure 6. Near α_c the amplitude behaves like $(\alpha-\alpha_c)^{\frac{1}{2}}$. It must be emphasised that in general there is no systematic way of assessing how near α must be to α_c for these results to hold, although a start has been made in certain special cases [Swinnerton-Dyer (1977)].

A general treatment of the Hopf bifurcation is given in the book by Marsden and McCracken (1976). Recently several sets of explicit rules have been given for identifying the super-critical case for sets of n equations. The most convenient one appears to be that of Poore (1976), which is not described by Marsden and

FIG 5 A B

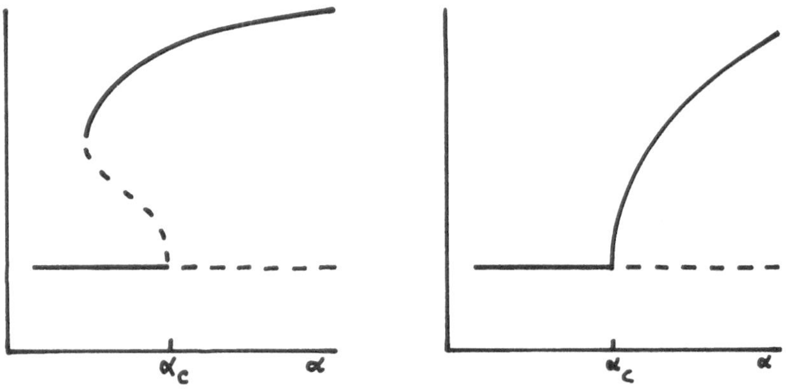
α_c
α
α_c
α

FIG 6

A: $\alpha < \alpha_c$
B: $\alpha > \alpha_c$
C: $\alpha < \alpha_c$
D: $\alpha > \alpha_c$

McCracken (1976) but is less laborious than the one that they describe. Its advantage lies in the fact that it is expressed in terms of the original variables, with no need to change variables prior to examining the non-linearities. If most of the n equations are linear, as in the applications that I shall consider, one gets the full advantage of this simplifying feature. Poore's treatment of this question is an elaborate one, but a more direct way to obtain his rule has been described by Allwright (1977a). I shall simply quote the essential result, in the same spirit as my quotation of the theorem of Hastings et al.

In applications to models incorporating the linear chain trick the critical parameter is the delay parameter a, or rather, with the convention that the critical parameter increases at the onset of instability, $\alpha = 1/a$. It is important that one should be able to evaluate α_c and the corresponding pure imaginary roots $\pm i\omega_o$ explicitly. Examples in which this is possible will be given in 4c and 5d.

Let the set of equations be defined as in (9), and let the corresponding Jacobi matrix be A, this matrix being evaluated at the critical value α_c of α. Let $\underset{\sim}{u}$, $\underset{\sim}{v}$ be the left and right eigenvectors of A with eigenvalue $+ i\omega_o$, normalised by setting $\underset{\sim}{u}.\underset{\sim}{v} = 1$. Let $\bar{\underset{\sim}{u}}$, $\bar{\underset{\sim}{v}}$ be the complex conjugates of these eigenvectors. Then super-critical or sub-critical bifurcation corresponds to positive or negative real part of the expression

$$\begin{aligned} \Phi = & -u_\ell \frac{\partial^3 g_\ell}{\partial x_j \partial x_m \partial x_s} v_j v_m v_s \\ & + 2u_\ell \frac{\partial^2 g_\ell}{\partial x_j \partial x_m} v_j (A^{-1})_{mr} \frac{\partial^2 g_r}{\partial x_p \partial x_q} v_p \bar{v}_q \\ & + u_\ell \frac{\partial^2 g_\ell}{\partial x_j \partial x_m} \bar{v}_j \left[(A-2i\omega_o I)^{-1}\right]_{mr} \frac{\partial^2 g_r}{\partial x_p \partial x_q} v_p v_q . \end{aligned} \tag{13}$$

In this equation all repeated indices imply summation. When the ordinary differential equations are preponderantly linear the summations are over relatively few values of the indices, since only a few second or third derivatives of the g_i are non-zero.

3d Numerical Integration

It may not be possible to apply either of the methods of the preceding sections. Also these methods each have their characteristic limitations. The theorem of Hastings et al. says nothing about the stability of the periodic solution. The Hopf method only applies "near" α_c. Neither method gives the form of the closed trajectory. They both allow an estimate of the period, $2\pi/\omega_0$, but again only close to α_c. So there are many reasons why one has to make use of numerical integration of the linear chain equations. If an apparently periodic solution is found in this way, within the accuracy of the numerical calculation, it can be assumed to be stable, since otherwise small numerical errors would cause the computed trajectory to drift away from the periodic one.

Numerical integration obliges one to look again at the relation between the original integro-differential equation and the equations of the linear chain. As mentioned in the introduction, to solve the first, for $t > 0$, one specifies $x(t)$ from $t = -\infty$ to $t = 0$. For the second one has n of the x_i given in the original formulation of the equation, so that their values at $t = 0$ can be assigned. Then there are p+1 additional x_i for which the initial values have to be inter-related for the solution to be the same as the solution of the integro-differential equation. The most straightforward case is one in which the initial state of the system is an equilibrium one, and it is assumed that at $t = 0$ a parameter of the model changes so that this equilibrium is destabilised, and the solution evolves in consequence. This implies that the x_i of the original problem have their equilibrium value at all $t < 0$, and that all the x_i of the linear chain problem have their equilibrium values at $t = 0$. There is no ambiguity concerning the shift from one problem to the other.

This will be the case for example if one seeks numerical solutions for a chemostat problem. The characteristic experimental method with the chemostat is to run the device until it settles into a steady state, then raise or lower the flow rate and examine the transient behaviour as the new steady state is approached. A similar situation prevails in the model of cyclical neutropenia to be studied in Chapter 6. The model assumes a stable steady state in normal health, with the onset of the disease associated with a change in one parameter, which destabilises the steady state and leads to periodic variation of the blood cell concentrations. While less natural in other problems, such as that of prey and predator populations, this procedure can still formally be used so long as in some parameter range the model has a stable equilibrium point.

Numerical integration of the differential equations may be inaccurate if a is large compared with other time scales of the model. However as indicated in Chapter 2 in these circumstances one expects solutions to be close to those of the instantaneous model in their qualitative behaviour. A Taylor expansion can be used when a is large. This requires knowledge of the low moments of the memory function,

$$\begin{aligned} \gamma_1 &= \int_0^\infty G(u)\, u du, \\ \gamma_2 &= \tfrac{1}{2}\int_0^\infty G(u)\, u^2 du, \\ &\quad \ldots\ldots\ldots\ldots \end{aligned} \tag{14}$$

since the delayed variable is written as

$$\begin{aligned} \bar{x} &= \int_{-\infty}^{t} x(\tau)\, G(t-\tau) d\tau = \int_0^\infty x(t-u)\, G(u) du \\ &= x(t) - \gamma_1 \frac{dx}{dt} + \gamma_2 \frac{d^2x}{dt^2} - \ldots \end{aligned} \tag{15}$$

However the Taylor expansion method must be used with great caution [Mazanov and Tognetti (1974), MacDonald (1975)].

Numerical integration is not a very efficient way of discriminating between the various situations portrayed in Figures 5 and 6, because in practice the region near α_c is very narrow, and the spiral trajectories are very tightly wound. A perturbation method may be available to supplement the Hopf analysis, as illustrated for the logistic model in the following chapter. This involves expansion in powers of $(\alpha-\alpha_c)^{\frac{1}{2}}$ in the super-critical case. Again this method has severe limitations. Alternative methods are becoming available (Langford 1977) for pursuing bifurcating solutions away from $\alpha = \alpha_c$.

LOGISTIC GROWTH OF A SINGLE SPECIES

4a Discrete Lag

To obtain a model in which the methods of Chapter 3 as well as those of Chapter 2 can be displayed, I shall look at a model even simpler than those discussed at the end of Chapter 2. A popular instantaneous model for the population growth of a single species, distributed homogeneously in space and having a finite upper limit to growth, is the logistic equation

$$\frac{dx}{dt} = rx(1-x/K). \tag{1}$$

This has two equilibrium points, $x = 0$, which is unstable, and $x = K$, which is stable. For small x, growth is exponential, then levels off, and eventually x tends asymptotically to K. The parameter r is a measure of the net reproductive rate of the species (birth rate - death rate) and K is the carrying capacity of the habitat for this species. Taking initial value x_1 the solution of equation (1) can be set down explicitly as

$$x(t) = \left\{\frac{1}{K} + \left[\frac{1}{x_1} - \frac{1}{K}\right] \exp(-rt)\right\}^{-1}. \tag{2}$$

Hutchinson (1948) introduced discrete lag into this model, in the second factor only, so that equation (1) becomes

$$\frac{dx}{dt} = rx(1-x(t-T)/K). \tag{3}$$

It is interesting to examine Hutchinson's paper, and to consider some implications of the particular form (3). Hutchinson takes as a general form of population growth equation

$$\frac{dx}{dt} = xf(x),$$

in which the function f(x) is the net reproductive rate per individual, that is to say the difference of birth and death rates. This is a natural formulation for mammal and bird populations, in which the adults are still around at the time of birth or hatching. Any complications of the population dynamics, which prevent the birth and death rates from simply being numbers, are put into the form of f(x), and this applies to lags as well. Hutchinson tacitly assumes, in writing (3), that birth and death rates are affected in the same way by the lag. He also assumes that it is appropriate to use discrete lag. The example he cites in support of the significance of lags is inconsistent with both assumptions. In his own words, "in Daphnia population oscillations can be set up which are in part due to the fact that the fecundity of the parthenogenetic female is determined not merely by the population density at a given time but also to the past densities to which it has been exposed".

For an insect population, in which all adults may have died before the new generation emerges, a delay in the birth terms, both linear and non-linear, would seem more appropriate. In the next section an example will be given of a model in which it is appropriate to think of the non-linear term which limits the population belonging solely to the death rate, but there discrete lag will be clearly wrong. These remarks are not intended to detract from the significance of Hutchinson's paper in drawing attention to the importance of lags. They merely indicate that the example to which this Chapter is devoted is primarily useful to illustrate techniques, rather than as a realistic representation of a population with a lag in its dynamics.

The analysis of the stability of the equilibrium point $x = K$ is trivial for the instantaneous model. The differential equation for $X = x-K$ is

$$\frac{dX}{dt} = r(X+K)(-X/K) \approx -rX,$$

and the stability equation is

$$\lambda + r = 0.$$

With discrete lag as in equation (3) the stability equation becomes, from (19) of Chapter 2,

$$\lambda + r\exp(-\lambda T) = 0. \qquad (4)$$

A transition to instability occurs when $\lambda = \pm i\omega_o$, with

$$i\omega_o + r\cos(\omega_o T) - ir\sin(\omega_o T) = 0, \qquad (5)$$

so that

$$\omega_o = r, \qquad T = \pi/2r. \qquad (6)$$

Thus the equilibrium point is stable so long as the lag T is below $\pi/2r$, while for greater lag it is unstable. The period of the solution at the critical lag is $2\pi/\omega_o$, so that the ratio of this period to the lag is 4.

According to a footnote in Hutchinson's paper, the earliest work on the nature of solutions beyond the critical value of T was carried out by Onsager, who showed that there must be fluctuations. Jones (1962) proved the existence of a periodic solution. He also presented results for a range of values of T, including values of the ratio T_{osc}/T rising from 4 to about 5.5 as T rises. It may be noted at this point that in the numerical work on chemostat models quoted in Chapter 2 this ratio seems again to have values about 4 or 5, so far as one can estimate it from the diagrams presented.

Some of Jones' results are reproduced in the book by May (1974a), in which equation (3) is used to fit the populations of the blowfly Lucilia cuprina in the experiment of Nicholson (1954). In this the blowfly were grown in liver, with overlapping populations and larvae taking about 11 days to become adults. May takes the lag to refer to this lapse of time, and does not analyse birth and death rates separately. He fixes rT = 2.1 from the amplitude of the oscillations. The corresponding period is 4.54T. From the observed period he finds T = 9 days.

Cunningham (1956) worked with a Taylor expansion in equation (3),

$$x(t-T) \quad = \quad x(t) - T\frac{dx}{dt} + \frac{1}{2}T^2\frac{d^2x}{dt^2} - \ . \ . \ .$$

To lowest order this gives (1) and to next order an equation for which the point x = K is always stable. The third order equation can be written

$$\frac{d^2x}{dt^2} - \frac{2dx}{dt} + \frac{2K}{rT^2x}\frac{dx}{dt} + \frac{2}{T^2}x = \frac{2K}{T^2}. \tag{7}$$

Here there are damping and anti-damping terms of different form so that, as for the van der Pol equation, oscillations may be expected. The stability polynomial corresponding to (7) and x = K can be obtained directly, or by using equation (4) with the Maclaurin expansion for $\exp(-\lambda T)$,

$$\exp(-\lambda T) \quad = \quad 1-\lambda T+\tfrac{1}{2}\lambda^2T^2- \ . \ . \ .$$

The polynomial is

$$\lambda^2\left(\frac{rT^2}{2}\right) + \lambda(1-rT) + r,$$

with roots

$$\lambda_1,\lambda_2 = \frac{1}{rT^2}\left\{(rT-1) \pm \sqrt{\left[(rT-1)^2 - 2r^2T^2\right]}\right\}. \tag{8}$$

Instability sets in at $T = 1/r$, which is rather a poor approximation to the correct result. For large rT the real part of the roots (8) behaves like $1/T$, so that the decay away from $x = K$ is slow for very long lags. This kind of behaviour was noted by Beddington and May (1975) in a model similar to the logistic, but with the lower, and unstable, equilibrium point at a finite value of x. They noted that the decay away from this point becomes slow at large T.

A much better approximation for the critical value of T can be found by converting (4) into a polynomial equation of order 3, not by extending the Maclaurin expansion by one more step, but by using the (2/2) Padé approximant to $\exp(-\lambda T)$,

$$\frac{1 - \frac{1}{2}\lambda T + \frac{1}{12}\lambda^2T^2}{1 + \frac{1}{2}\lambda T + \frac{1}{12}\lambda^2T^2}, \tag{9}$$

which is roughly equivalent to extending the Maclaurin expansion up to $(\lambda T)^5$. The resulting stability equation is

$$\lambda^3 + \lambda^2\left(\frac{6}{T} + r\right) + \lambda\left(\frac{12}{T^2} - \frac{6r}{T}\right) + \frac{12r}{T^2} = 0. \tag{10}$$

The Routh-Hurwitz determinant Δ_2 is

$$-\frac{6}{T^3}\left((rT)^2 + 6rT - 12\right),$$

and setting this equal to zero gives $T = 1.58/r$.

The effectiveness of this method depends on the fact that the roots relevant to the stability determination are those of lowest modulus. Re-examining equation (4) with

$$\lambda = \alpha + i\beta,$$

$$|\lambda| = \sqrt{\alpha^2 + \beta^2},$$

one can readily see that

$$\frac{\alpha}{r} = -e^{-\alpha T} \cos(\beta T),$$

$$\frac{\beta}{r} = e^{-\alpha T} \sin(\beta T),$$

so that

$$\frac{|\lambda|}{r} = e^{-\alpha T}. \tag{11}$$

Any root with positive real part has $|\lambda| < r$, while any root with negative real part has $|\lambda| > r$.

4b Distributed Lag in a Model of a Self-poisoning Population

The first work using a logistic equation with distributed lag was by Volterra (1934), with extensions by Kostitzin (1937). In the 1930's, largely because of the influence of Volterra's work, many experiments were performed with laboratory populations of one or two species of small organisms with short generation times. Attempts to apply logistic models to these experiments were often frustrated because populations died out instead of attaining a sustained steady value. It was realised that one cause of this was the pollution of the closed environment by waste products and dead organisms. Thus Volterra set out to examine a cumulative effect in the death rate of a species, dependent on the population at all times from the start of the experiment.

A somewhat analogous situation can occur for a parasite which completes its life cycle within the same host, and does not kill the host. Immunological resistance by the host depends on exposure to the parasite population. To quote Michel (1969), "characteristically the increase is exponential during early stages of infection when the host offers an ideal environment. Subsequently, when the host becomes resistant and represents a less suitable environment, the rate of increase declines to zero and the population then rapidly decreases."

An appropriate model is an integro-differential equation

$$\frac{dN}{dt} = rN \left\{ 1 - \frac{N}{K} - \int_0^t N(\tau)\ G(t-\tau)\ d\tau \right\}, \tag{12}$$

in which the quadratic instantaneous term is accompanied by a pollution term. It should be noted that in this context it is appropriate to take the integral from $t = 0$. This initial time represents the start of the experiment, or the time at which the naive host ingests parasites. It is therefore possible to use the simplest of all memory functions, G = constant. To illustrate the nature of the results, ignore the instantaneous quadratic term and use

$$\frac{dN}{dt} = N\left\{r-c\int_0^t N(t)\,dt\right\}.$$

Writing

$$M(t) = \int_0^t N(t)\,dt,$$

this is the same as the differential equation

$$\frac{d^2M}{dt^2} = (r-cM)\frac{dM}{dt},$$

equivalent to a modified logistic equation for M,

$$\frac{dM}{dt} = rM - \frac{cM^2}{2} + N_o. \tag{13}$$

The solution is

$$M(t) = \frac{\alpha\beta(e^{\gamma t}-1)}{(\alpha+\beta e^{\gamma t})},$$

giving

$$N(t) = \frac{N_o(\alpha+\beta)^2\, e^{\gamma t}}{(\alpha+\beta e^{\gamma t})^2}. \tag{14}$$

Here α and β are the roots of the equation

$$M^2 - \frac{2r}{c}M - \frac{2N_o}{c} = 0,$$

so that α is the value approached by M at large values of t. The exponent γ is defined to be

$$\tfrac{1}{2}c(\alpha+\beta).$$

The population (14) rises to a maximum and then falls off as $\exp(-\gamma t)$.

4c Linear Chain Calculations

I shall now examine the logistic model with distributed lag, with the integration running from $t = -\infty$, and with the memory function $G_a^p(u)$. The stability equation, from (32) of Chapter 2, is

$$\lambda + r\left(\frac{a}{a+\lambda}\right)^{p+1}. \tag{15}$$

For $p = 0$ this implies stability for all a, but for $p \geq 1$ instability is attained if a is small enough. For $p = 1$, equation (15) is equivalent, at finite a, to

$$\lambda^3 + 2a\lambda^2 + a^2\lambda + a^2 r = 0.$$

The Routh-Hurwitz determinant Δ_2 is $2a^3 - ra^2$, so that the real part of the complex pair of roots changes sign, at $a = r/2$. This corresponds to a critical value of the average lag, $\bar{T} = 2/a$, of $4/r$. This is substantially larger than the critical value of T in the discrete lag case.

Clearly this would be a poor approximation to the discrete lag result, as compared with the use of the Padé approximant yielding an equation, (10), of the same order in λ. However that is not what is intended. Distributed lag is liable to be more realistic than discrete lag, and it is worth noting that it tends to require a longer average lag to yield instability.

For $p = 2$ the determinant Δ_3 is easily evaluated to obtain the critical value of the average lag as

$$\bar{T} = 3/a = 8/3r,$$

which is appreciably nearer to the discrete lag value. As p increases beyond 2 this method rapidly becomes painful, and it is more profitable to examine the condition for a pair of pure imaginary roots. The value of the pure imaginary roots also yields an estimate of T_{osc}. I shall return to the question of whether there is a periodic solution.

One must find ω_o such that

$$i\omega_o(i\omega_o+a)^{p+1} = -a^{p+1}r, \tag{16}$$

which can be written as

$$(\cos\phi+i\sin\phi)^{p+1} = \left(-\frac{r}{a}\right)(\cos\phi)^{p+2}/i\sin\phi, \tag{17}$$

with

$$\tan\phi = \frac{\omega_o}{a}. \tag{18}$$

Now equation (17) gives

$$\phi = \frac{\pi}{2(p+1)},$$

and

$$a\sin\phi = r(\cos\phi)^{p+2}.$$

So, for fixed r, instability sets in at

$$a = r\left\{\cos\left(\frac{\pi}{2(p+1)}\right)\right\}^{p+2} \Big/ \sin\left(\frac{\pi}{2(p+1)}\right), \tag{19}$$

FIG 7

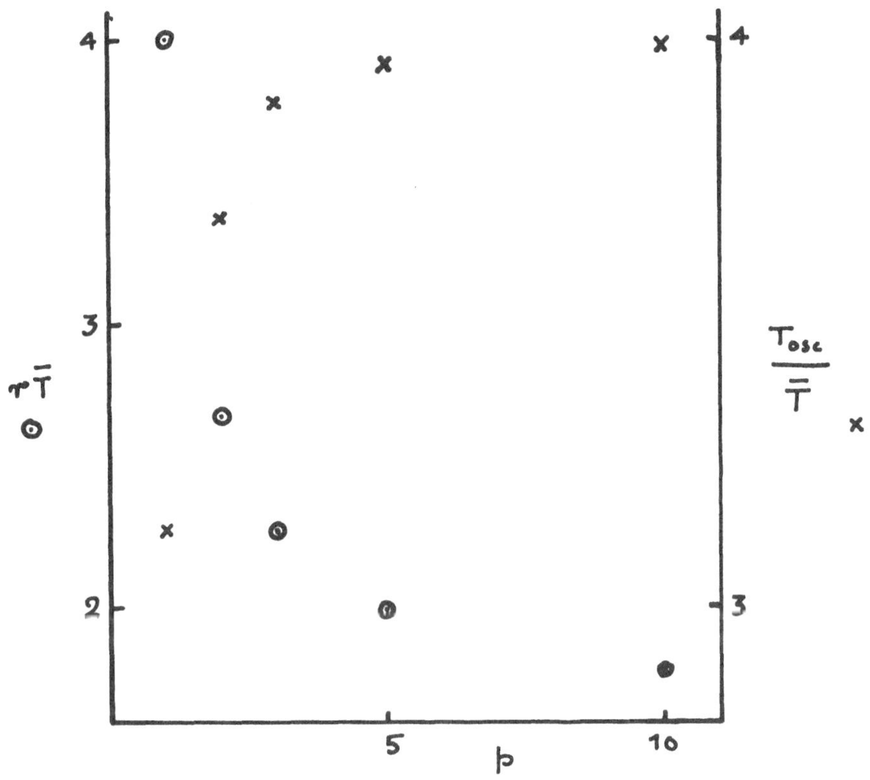

and the corresponding frequency is

$$\omega_o = a\tan\left(\frac{\pi}{2(p+1)}\right) = r\left\{\cos\left(\frac{\pi}{2(p+1)}\right)\right\}^{p+1}. \tag{20}$$

Hence

$$T_{osc}/\bar{T} = \frac{2\pi}{\omega_o}\Big/\frac{p+1}{a} = \frac{2\pi}{p+1}\cot\left(\frac{\pi}{2(p+1)}\right). \tag{21}$$

For large p and a, fixing $\bar{T}$ as constant, one recovers the discrete lag result, as expected. The general nature of these results is displayed in Figure 7. The critical $\bar{T}$ is always greater than the critical T of the discrete case, and the ratio $T_{osc}/\bar{T}$ is always smaller than the ratio T_{osc}/T of the discrete case. As p increases, the discrete case results are approached.

4d Hopf and H.T.W. Methods

In 4c the distributed lag with memory function $G_a^p(u)$ did not need to be treated by the linear chain trick. In this section I shall return to the use of the expanded set of equations, which for this problem is the p+2 dimension set

$$\frac{dx_1}{dt} = rx_1\,(1-x_{p+2}/K) = g_1\,(x_1\ .\ .\ .\ x_{p+2}),$$

$$\frac{dx_s}{dt} = a(x_{s-1}-x_s) = g_s(x_1\ .\ .\ .\ x_{p+2}), \qquad (22)$$

$$2 \leqslant s \leqslant p+2$$

As indicated in 4c the change from stable to unstable equilibrium point occurs as the real part of a pair of complex conjugate roots of the stability equation changes from negative, for large a, to positive, for small a, at the critical value (19) of a. In the p+2 dimension phase space of the variables in (22) the trajectories spiral in towards (K,K, . . ., K) for larger a, and spiral out from this point for smaller a. Just at the critical value of a this point is marginally stable, and is surrounded by closed trajectories of period $2\pi/\omega_0$ and arbitrary amplitude.

The questions to be examined now are whether, just below the critical a, the outward spirals near the equilibrium tend to a stable periodic trajectory, and whether, for any a below the critical value, there is a periodic solution of (22). The first is treated using Poore's version of the Hopf bifurcation method, the second by applying the theorem of Hastings, Tyson and Webster. From 4c it is known that the critical a and the corresponding frequency ω_0 can be evaluated explicitly for any p. Also from (22) it is obvious that only two non-zero second derivatives enter the Poore algorithm. So the Hopf investigation should be straightforward.

No g_k has non-zero third derivatives, and the only non-zero second derivatives are

$$\frac{\partial^2 g_1}{\partial x_1 \partial x_{p+2}} = \frac{\partial^2 g_1}{\partial x_{p+2} \partial x_1} = \frac{-r}{K}\ . \qquad (23)$$

Consequently, applying (13) of Chapter 3, there is a stable periodic solution of

(22) just below critical a if the real part of the following expression is positive:

$$\Phi = 2\left(\frac{r}{K}\right)^2 u_1 \left[v_1(A^{-1})_{p+2,1} + v_{p+2}(A^{-1})_{1,1}\right]\left(v_1\bar{v}_{p+2} + \bar{v}_1 v_{p+2}\right)$$

$$+ 2\left(\frac{r}{K}\right)^2 u_1 \left\{\bar{v}_1 \left[(A-2i\omega_o I)^{-1}\right]_{p+2,1} + \bar{v}_{p+2}\left[(A-2i\omega_o I)^{-1}\right]_{1,1}\right\} v_1 v_{p+2} \qquad (24)$$

Here A is the (p+2)×(p+2) Jacobian matrix of (22),

$$\begin{pmatrix} 0 & 0 & . & . & . & -r \\ a & -a & 0 & . & . & 0 \\ 0 & a & -a & 0 & . & 0 \\ . & . & . & . & . & . \\ . & . & . & 0 & a & -a \end{pmatrix},$$

in which a takes its critical value a_c. It is convenient to adopt the notation that $a_c = \beta r$, the corresponding $\omega_o = \alpha r$, and the ratio $\omega_o/a = \alpha/\beta = \gamma$. Then by equations (19) and (20) these new parameters are given by

$$\alpha = \left\{\cos\left(\frac{\pi}{2(p+1)}\right)\right\}^{p+1},$$

$$\beta = \alpha \cot\left\{\frac{\pi}{2(p+1)}\right\},$$

$$\gamma = \tan\left\{\frac{\pi}{2(p+1)}\right\}.$$

The eigenvectors $\underset{\sim}{u}$ and $\underset{\sim}{v}$ for A, with eigenvalue $+i\omega_o$, are easily set down. In particular for the right eigenvector the result

$$-v_{p+2} = i\alpha v_1$$

means that the first line of (24) vanishes. The second line reduces to $2(r/K)^2$ times

$$\left\{\frac{1+i\gamma}{1+i\gamma(p+2)}\right\}(-i\alpha)\left\{\left[(A-2i\omega_o I)^{-1}\right]_{p+2,1} + i\alpha\left[(A-2i\omega_o I)^{-1}\right]_{1,1}\right\},$$

in which the first factor comes from the normalisation condition for the eigenvectors, $\underset{\sim}{u}.\underset{\sim}{v} = 1$. Evaluating the necessary elements of the inverse matrices leads to the compact expression

$$\frac{(1+i\gamma)i\alpha\,(1+i\alpha\,[1+2i\gamma]^{p+1})}{(1+i\gamma(p+2))\,(1+2i\alpha\,[1+2i\gamma]^{p+1})}. \qquad (25)$$

Figure 8 shows the real part of (25) for p = 1 to p = 10. One can also easily verify that the real part of (25) tends to a positive limiting value as p becomes large.

The conclusion is that this form of distributed lag in the logistic model leads to periodic behaviour just below the critical value βr of a. In his book Cushing (1978) presents a treatment of the case p = 1, using a related set of ordinary differential equations, and employing the method of Poore. In one of his

FIG 8

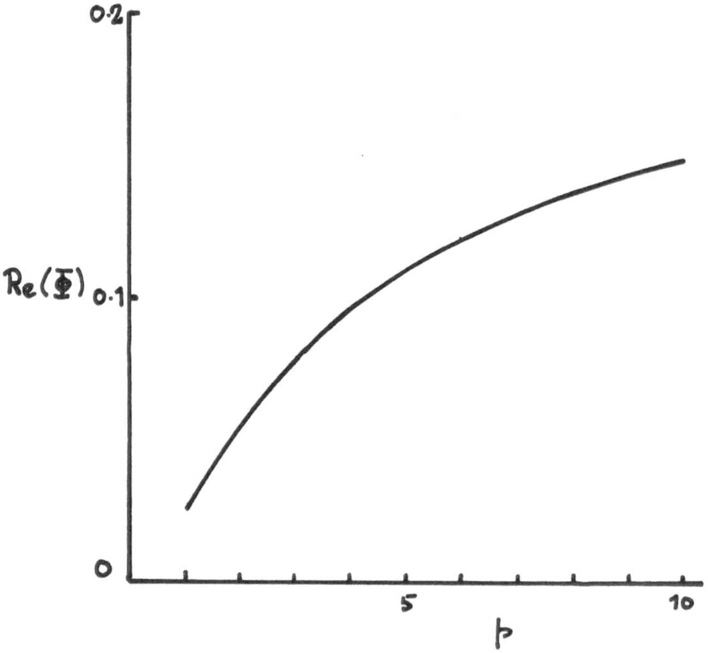

papers [Cushing (1976b)] he gives another way of exploring bifurcations in the logistic model, applicable to more general memory functions.

Returning to equations (22), it is straightforward to show that the conditions of the theorem of Hastings et al. as given in equations (10), (11) and (12) of Chapter 3, are satisfied. Equations (22) have a unique positive equilibrium point for positive x_k. The stability equation,

$$Q(\lambda) \quad = \quad \lambda(a+\lambda)^{p+1} + ra^{p+1} \quad = \quad 0,$$

has no repeated root. Any repeated root must also be a root of

$$Q'(\lambda) = (a+\lambda)^p (a+\lambda(p+2)) = 0.$$

These last two equations have no common root for finite non-zero a. So all the conditions of the theorem apply, and there is at least one non-constant periodic solution for a less than the critical value βr.

Some authors [Goel et al. (1971), Montroll (1972), Goel and Richter-Dyn (1974)] have used a generalised logistic equation

$$\frac{dx}{dt} = rx\left[1 - (x/K)^v\right]/v.$$

This gives equation (1) for v = 1, and the Gompertz equation

$$\frac{dx}{dt} = -rx \ln(x/K)$$

as v tends to zero. Inserting a distributed delay in the second factor, with the memory function $G_a^p(u)$, the stability analysis goes through unchanged, and consequently the Hopf bifurcation algorithm can be set up with the same eigenvectors and matrix inverses. However for $v \neq 1$ there are three non-zero second derivatives and four non-zero third derivatives, and the calculations become considerably more tedious. On the other hand the application of the theorem of Hastings, Tyson and Webster is just as direct as in the case v = 1, and so it is clear that this form of distributed delay leads to a periodic solution so long as the equilibrium point is unstable.

4e Constant Harvesting of a Population in the Presence of Lag

A more interesting generalisation of the logistic equation (1) includes harvesting at a constant rate,

$$\frac{dx}{dt} = rx(1-x/K) - C. \qquad (1a)$$

The equilibrium points, which for C = 0 are at x = 0 and x = K, are in general given by the roots of the equation

$$x^2 - Kx + KC/r = 0.$$

A fisheries policy may be guided, in a crude approximation, by the attempt to keep C as high as possible while remaining at the stable upper equilibrium point,

$$x_o = \frac{K}{2}\left[1 + \sqrt{(1-4C/Kr)}\right]. \qquad (2b)$$

So one attempts to set

$$C \lesssim Kr/4$$

so that

$$x_o \gtrsim K/2.$$

For practical purposes one must leave a margin of error between C and Kr/4, since one can neither completely control C nor precisely predict K and r. On the other hand it is easily seen that one is in little danger, for C close to Kr/4, of instability due to time delay. All the stability results of this Chapter go through simply by multiplying r by $\sqrt{(1-4C/Kr)}$.

The minimum mean lag for instability, which from equation (19) is proportional to

$$1/r \sqrt{(1-4C/Kr)},$$

becomes very large as $C \to Kr/4$. So it is likely that realistic lags will either not exceed this critical value, or if they do will give very slow oscillations, by the result (21).

In fact a system which is almost unstable, and so has a sluggish return to its equilibrium point when disturbed, is not likely to acquire a dangerous overshoot in the presence of lag.

4f <u>The Poincaré-Lindstedt Method for Discrete Lag</u>

Morris (1976) presents a perturbation method for use with the logistic model, both for discrete lag and for distributed lag with memory function $G_a(u)$. This method supplements the methods of Chapter 3 since it clarifies the behaviour of the periodic solution near the critical lag. The method can also be used for sets of two or more equations with one lag [Morris (1978)].

When dealing with discrete lag, in line with the convention used by Jones (1962), Morris takes r = K = 1, and changes the variable to y = x-1, so that the equilibrium point of interest is y = 0. The equation (1) now becomes

$$\frac{dy}{dt} = -y(t-T)(1+y(t)). \tag{1b}$$

The variable y is assumed small, so that the method only applies when the population oscillations have small amplitude compared to K. From the numerical results of Jones this implies keeping T within $\sim$ 0.03 of the critical value $\pi/2$. An essential feature of the method is the choice of expansion parameter. Since it is known that below a critical value β of T there is no non-constant periodic solution, while above T = β there is such a solution, the assumption is made that the amplitude and frequency are analytic not in (T-β) but in $\varepsilon = (T-\beta)^{\frac{1}{2}}$, and they are expanded as

$$\begin{aligned} y(t) &= \varepsilon Y(t) = \varepsilon\left[Y_o(t) + \varepsilon Y_1(t) + \ldots\right], \\ \omega(\varepsilon) &= \omega_o + \varepsilon\omega_1 + \varepsilon^2\omega_2 + \ldots \end{aligned} \tag{27}$$

It is convenient to use functions of known period, so that a new independent variable is defined,

$$s = \omega(\varepsilon)t,$$

and finally one writes

$$Y(t) = X(s),$$

so that X(s) is periodic with period 2π. The perturbation expansion is now applied in the equation

$$\omega(\varepsilon)\frac{dX}{ds} = -X(s-\alpha(\varepsilon))(1+\varepsilon X), \tag{28}$$

in which

$$\alpha(\varepsilon) = \omega(\varepsilon)T = \omega(\varepsilon)(\beta+\varepsilon^2) = \omega_o\beta + \varepsilon\beta\omega_1 + \varepsilon^2(\omega_o+\beta\omega_2) + \ldots \tag{29}$$

The delayed variable $X(s-\alpha)$ is written as

$$X(s-\alpha(\varepsilon)) = X_o(s,\alpha) + \varepsilon X_1(s,\alpha) + \ldots \tag{30}$$

corresponding to the expansion in (27), which is equivalent to

$$X(s) = X_o(s) + \varepsilon X_1(s) + \ldots \tag{31}$$

To take account of the expansion (29) each term in the expansion of $X(s-\alpha)$ is itself expanded as a Taylor series,

$$X_j(s,\alpha) = X_{j,\beta\omega_o} + \varepsilon\beta\omega_1\frac{dX_{j,\beta\omega_o}}{ds} + \ldots \tag{32}$$

The overall expansion looks rather formidable at this stage, but there are some features which help to keep the first few terms manageable. The initial condition $X_j(0) = 0$ will be used at all orders j.

I. Terms of order ε^0

These give a simple equation, which determines β, ω_o and the phase relation between X_o and $X_{o,\beta\omega_o}$. This equation is

$$\omega_o \frac{dX_o}{ds} = -X_{o,\beta\omega_o}. \tag{33}$$

Using $X_o(0)=0$, this gives

$$\omega_o = 1, \quad \beta = \pi/2.$$
$$X_o = B_o \sin(s) \quad X_{o,\beta\omega_o} = -B_o \cos(s). \tag{34}$$

The coefficient B_o is at this stage undetermined.

II. Terms of order ε^1.

The most important feature of the method that emerges at first order is the elimination of secular terms. The equation is found to have two parts that contribute terms with cos(s) or sin(s) behaviour to X_1. These can only be consistent if one of them gives secular terms such as ssin(s), or if this part is set equal to zero. Secular terms must be avoided, since there is known to be a periodic solution, so the alternative must be adopted. In this order this means that $\omega_o = 0$. This has a considerable simplifying effect in higher orders. The equation is

$$\omega_o \frac{dX_1}{ds} + \omega_1 \frac{dX_o}{ds} = -\left(X_{1,\beta\omega_o} + X_o X_{o,\beta\omega_o} + \beta\omega_1 \frac{dX_{o,\beta\omega_o}}{ds} \right)$$

or, using (34) and in particular simplifying the notation by setting $\beta\omega_o = \beta$,

$$\frac{dX_1}{ds} + X_{1,\beta} = \frac{B_o^2}{2} \sin(2s) - \omega_1 \left(\frac{dX_o}{ds} + \beta \frac{dX_{o,\beta}}{ds} \right). \tag{35}$$

This reduces by the above argument to

$$\frac{dX_1}{ds} + X_{1,\beta} = \frac{B_o^2}{2} \sin(2s). \tag{36}$$

Applying the condition $X_1(0) = 0$, this has solution

$$X_1 = B_1 \sin(s) - \frac{B_o^2}{10} \sin(2s). \tag{37}$$

III. Terms of order ε^2.

B_o is still undetermined. In fact B_o and ω_2 are determined by the elimination of secular terms from the equation of order ε^2, which is

$$\frac{dX_2}{ds} + \omega_2 \frac{dX_o}{ds} = -X_{2,\beta} + (\omega_o + \beta\omega_2) \frac{dX_{o,\beta}}{ds} - X_{o,\beta} X_1 - X_{1,\beta} X_o. \tag{38}$$

By inserting the results for X_o and X_1 this becomes

$$\frac{dX_2}{ds} + X_{2,\beta} = \frac{-B_o^3}{20}\cos(3s) - \frac{3B_o^3}{20}\sin(3s) + \frac{B_o^3}{5}\cos(2s) + B_oB_1\sin(2s)$$

$$+ \cos(s)\left[-B_o\omega_2 - \frac{3B_o^3}{20}\right] + \sin(s)\left[B_o(\omega_o+\beta\omega_2) + \frac{B_o^3}{20}\right].$$

Once again the two terms on the left give all the sin(s) or cos(s) behaviour that can be present in X_2 unless there are secular terms, so the coefficients of sin(s) and cos(s) on the right must be set equal to zero. This provides two equations from which to determine ω_2 and B_o, which turn out to be

$$\omega_2 = -0.81, \qquad B_o = 2.32$$

Morris also gives the equation of order ε^3, and from it obtains ω_3 and B_1. Because of the very narrow range of T over which the amplitude of the periodic solution remains small compared with K, this method is not a practical one for computation in this problem. It is better thought of as a theoretical method parallel to that of Hopf bifurcation. As already mentioned, Morris has applied the method to $G_a^p(u)$, and there is no reason why it should not go through for higher p.

The feature of this method that makes it possible to eliminate secular terms and so approximate the periodic solution in a well-defined manner, is the simultaneous expansion of both frequency and amplitude in terms of ε. One point should be noted, which is not explicitly pointed out by Morris, concerning the elimination of secular terms from order ε^2. The equations that result are for ω_2 and B_o^2, so they imply that the expansion is only meaningful if B_o^2 is found to be equal to a positive quantity. So the method contains a test for the existence of the periodic solution. This quantity B_o^2 turns out to contain a sum of terms linear in the third derivatives of the non-linear terms in the original equation, and of terms quadratic in the second derivatives of these non-linear terms, as in the case of Poore's test for super-critical Hopf bifurcation. Connections between these methods, applied to the logistic equation, are discussed in a recent paper by Kazarinoff et al. (1978).

4g An Epidemic Model Related to the Logistic Equation

K.L. Cooke (1978) has recently examined an epidemic model leading to the delay-differential equation

$$\frac{dy}{dt} = by(t-T)(1-y(t)) - cy(t). \tag{39}$$

This affords an example in which lag is biologically meaningful, but has no effect on the stability properties of the equilibrium points.

The variable y(t) is the infected human population. It is assumed that the total human population is constant, and this is scaled = 1. So if x(t) is the population of uninfected humans,

$$x(t) + y(t) = 1.$$

The disease is transmitted to man by an insect vector, assumed to have a large and constant population, and by man to that vector. Within the vector there is an incubation period T before the disease agent can infect a human. So the population of vectors capable of infecting man is

$$z(t) = dy(t-T).$$

Infection of man is assumed to proceed at a rate proportional to encounters between uninfected humans and vectors capable of transmitting the disease,

$$ex(t).dy(t-T) = by(t-T)\{1-y(t)\},$$

and recovery to proceed exponentially at a rate c. Infection leads neither to death, immunity or isolation. These assumptions then give the equation (39).

For all b, c, y = 0 is an equilibrium point. For c < b, y = 1 - c/b is also an equilibrium point. At y = 0 the linearised equation is

$$\frac{dy}{dt} = by(t-T) - cy(t). \tag{40}$$

At the other equilibrium point, writing

$$y(t) = \left\{1 - \frac{c}{b}\right\}\left\{1 + v(t)\right\},$$

the linearised equation is

$$\frac{dv}{dt} = cv(t-T) - bv(t). \tag{41}$$

So all conclusions regarding (40) and (41) are equivalent, with b and c interchanged.

The local stability of the point $y = 0$, for $b < c$, is readily established. The stability equation is

$$\lambda + b \exp(-\lambda T) - c = 0.$$

This can obviously not have a solution $\lambda = 0$, nor can it have a pure imaginary solution $\lambda = i\omega_o$ since this would require

$$\cos(\omega_o T) = c/b > 1.$$

When $T = 0$, λ is less than zero, so the real part of λ must remain negative for all T.

The greater part of Cooke's paper is devoted to the much more difficult task of proving that the global stability properties are also independent of T. He shows that any y(t) starting in the range 0 to 1 will tend to 0 if c exceeds b, and to 1 if b exceeds c. Thus there is no question of oscillations being sustained in this model.

BIOCHEMICAL OSCILLATOR MODEL

5a The Goodwin Model

Sequences of biochemical reactions in which the end product inhibits the first reaction are not uncommon. They represent a rather simple way of controlling the amount of the end product formed, and arguments can be given [Savageau (1976)] to show that this particularly simple feedback loop has advantages over other possible modes of control. One rarely has sufficiently detailed data to attempt to build a model for a specific sequence of this kind, and quite a lot of work has been in terms of a schematic general model due to Goodwin (1965). In this model all the reactions have linear kinetics, except for the inhibition process. Goodwin suggested that in his model the negative feedback should lead to oscillations. Early work was mainly numerical, but in the last few years a considerable degree of progress has been made in the analytical study of periodic solutions of the equations of the model. For a comprehensive review see Tyson and Othmer (1978).

As outlined in section 1d a case can be made for including lags in a model of this type. In this chapter I shall apply the inequality of section 2e to obtain a result excluding instability of the equilibrium for an important case of the Goodwin model, for any reasonable memory function. I shall also apply the linear chain trick so as to make available the methods of Chapter 3 in the presence of a distributed lag, for the memory functions $G_a^p(u)$.

The Goodwin model is expressed in terms of the following set of equations, in which the variables are the concentrations of various chemical species,

$$\frac{dx_1}{dt} = \frac{k_1}{1+x_n^\rho} - b_1x_1,$$
$$\frac{dx_i}{dt} = k_ix_{i-1} - b_ix_i, \quad i = 2,3, \ldots n. \tag{1}$$

The parameter ρ is a positive integer, the Hill coefficient, the significance of which is that ρ molecules of the end product combine with an enzyme which catalyses the first reaction. This combination gives an inactive product, so that by switching off the catalysis the first reaction is inhibited. There is a qualitative distinction between the cases $\rho = 1$ and $\rho \geqslant 2$, the second involving a co-operative inhibition process. Cases are known in which $\rho > 1$ is required to fit the data; also cases with n as large as 8 or 9 are known [Tyson and Othmer (1978)].

It is now well known that there is a striking difference between the solutions of (1) with $\rho = 1$ and those with $\rho \geqslant 2$. For $\rho \geqslant 2$ it is always possible, by taking large enough n, to have unstable equilibrium in some range of values of the parameters. Then there is a super-critical Hopf bifurcation. This has been proved by the method of Poore [MacDonald (1977b), Mees and Rapp (1978)] and by an alternative method [personal communication from J.J. Tyson]. In my work on this problem a special case is used in which all the parameters in the linear terms of (1) are set equal; it is then possible to set down a_c and ω_o explicitly, in a manner similar to that employed in 4d. The calculation by Mees and Rapp (1978) allows more general parameter values by first taking n to be large.

As mentioned in 3b, the method of Hastings et al. (1977) was devised with this model in mind. They show that whenever the equilibrium point of (1) is unstable there is at least one non-constant periodic solution. On the other hand, for $\rho = 1$ Allwright (1977b) has shown that the locally stable equilibrium point is in fact a global attractor, so that there is no non-constant periodic solution. His result goes through also if there is a discrete lag in the non-linear term of equation (1).

Probably the earliest comment on lag in this model was made by Landahl (1969). He considered discrete lag in the variable x_n in the non-linear term of the first equation. He found that for $\rho = 1$ and any n the equilibrium point is stable for all T. His method was essentially the use of the inequality discussed in 2e.

The results of the following two sections come from my paper [MacDonald (1977c)].

5b Necessary Condition for Instability

The stability equation for this model is

$$(\lambda+b_1)(\ .\ .\ .\ )(\lambda+b_n) + \beta k_1\ .\ .\ .\ k_n = 0. \tag{2}$$

I shall only look at cases that are stable without lag, so that all roots of (2) are taken to have negative real part. The parameter β in (2) is defined by

$$\beta = \frac{\rho x^{\rho-1}}{(1+x^{\rho})^2},$$

where x is the equilibrium value of x_n, satisfying

$$\frac{b_1 \ . \ . \ . \ b_n}{k_1 \ . \ . \ . \ k_n} \quad x \quad = \quad \frac{1}{1+x^\rho} \ .$$

Hence

$$\frac{b_1 \ . \ . \ . \ b_n}{k_1 \ . \ . \ . \ k_n \beta} \quad = \quad \frac{x^\rho + 1}{\rho x^\rho} \quad > \quad \frac{1}{\rho} \ . \tag{3}$$

The stability of the equilibrium point when a distributed lag, with memory function G(u), is inserted in any off-diagonal term in the equations (1) is now determined by the equation

$$(\lambda+b_1)(\ . \ . \ . \ )(\lambda+b_n) + \beta k_1 \ . \ . \ . \ k_n \ F(\lambda) \quad = \quad 0, \tag{4}$$

where as usual $F(\lambda)$ is the Laplace transform of G(u). Since $F(\lambda)$ is a positive function for real positive λ, the fact that (2) has no real positive root means that neither does (4). So the onset of instability must be by way of a pair of pure imaginary roots $\pm i\omega_o$. For real positive ω_o there must be a root of

$$\frac{(i\omega_o+b_1)(\ . \ . \ . \ )(i\omega_o+b_n)}{\beta k_1 \ . \ . \ . \ k_n} \quad = \quad -F(i\omega_o) \ . \tag{5}$$

Now by the argument given in 2e the modulus of the right hand side of (5) must be no larger than 1. By equation (3) the modulus of the left hand side must be larger than $1/\rho$. So for $\rho = 1$ the equilibrium point remains stable in the presence of an off-diagonal lag. For $\rho \geqslant 2$ the stability analysis must be pursued further. It is clear that any number of the off-diagonal terms can have a lag without changing this result, the right hand side of (5) becoming the product of a number of Laplace transforms all with modulus no greater than 1.

5c Expanding the Set of Equations

The linear chain trick introduces p+1 new linear equations of the type that appear in (1), identifying k_j and b_j with a for the new variables x_j. I have therefore to verify that the expanded set of equations is still of the form (1), in which case all the known results for the Goodwin model can be used to cover the case of distributed lags with the memory function $G_a^p(u)$. I shall do this only for n = 2, but the extension is obvious. I only consider lags in off-diagonal terms. With lag in the non-linear term in the first equation, the equations

$$\frac{dx}{dt} = \frac{k_1}{1+z^{\rho}} - b_1 x$$

$$\frac{dy}{dt} = k_2 x - b_2 y \tag{6}$$

$$\frac{dz}{dt} = a(w-z)$$

$$\cdot \; \cdot \; \cdot \; \cdot \; \cdot \; \cdot \; \cdot \; \cdot \; \cdot$$

$$\frac{dr}{dt} = a(y-r)$$

become of the form (1) with the identification

$$\begin{array}{lll} x_1 = x & b_j = k_j = a & j \geq 3 \\ x_2 = y & b_1 = b_1 & \\ x_3 = r & k_1 = k_1 & \\ \cdot \; \cdot \; \cdot \; \cdot \; \cdot & b_2 = b_2 & \\ x_n = z & k_2 = k_2 & \end{array} \tag{7}$$

The equations obtained by putting lag in the off-diagonal term in the second equation,

$$\frac{dx}{dt} = \frac{k_1}{1+y^{\rho}} - b_1 x$$

$$\frac{dy}{dt} = k_2 z - b_2 y \tag{8}$$

$$\frac{dz}{dt} = a(w-z)$$

$$\cdot \; \cdot \; \cdot \; \cdot \; \cdot \; \cdot \; \cdot \; \cdot \; \cdot \; \cdot$$

$$\frac{dr}{dt} = a(x-r)$$

become of the form (1) with the identification

$$\begin{array}{lll} x_1 = x & b_j = k_j = a, & j = 2, \ldots n-1 \\ x_2 = r & b_1 = b_1 & \\ \cdot \; \cdot \; \cdot \; \cdot \; \cdot & k_1 = k_1 & \\ x_{n-1} = z & b_n = b_2 & \\ x_n = y & k_n = k_2 & \end{array} \tag{9}$$

So the effect is indeed to lengthen the chain of equations.

Now assume that there are lags in a number of off-diagonal terms involving concentrations x_i, with p_i the order of the lag in each case. If one defines

$$P = \sum_i (p_i+1),$$

the Goodwin model has its dimension increased from n to n' = n+P. All the results quoted in 5a hold. With $\rho = 1$ any number of these distributed lags, of arbitrarily high order, can be put into off-diagonal terms, and the equilibrium point remains a global attractor. When $\rho \geq 2$, the equilibrium point can become unstable if sufficient lags are added, or one lag is added of sufficiently high order, to make n' attain the required value. In this case the Hopf and H.T.W. results hold.

The conclusion is that the one-loop Goodwin model (1) requires cooperative inhibition to make oscillations possible, even in the presence of lags in off-diagonal terms. The effect of lags, at any rate with memory function $G_a^p(u)$, is analogous to lengthening the chain of equations, and so for $\rho \geq 2$ they can contribute to the onset of oscillations.

Now there is evidence [Mees and Rapp (1978)] that modifying this kind of model to include parallel feedback loops can make it more prone to instability of the equilibrium point. So one may ask, what about the effect of diagonal lag? According to the argument in 2d this corresponds to adding a loop if the linear chain trick is used, or at any rate to converting a trivial 1-loop to a loop of length p+2. However it is difficult to see what meaning could be attached to a lag in the degradation of one of the chemical species. Also in the Goodwin model diagonal lags lead to negative concentrations. This is easily seen in the case

$$\frac{dx}{dt} = \frac{k_1}{1+y} - b_1 x,$$

$$\frac{dy}{dt} = k_2 x - b_2 y(t-T),$$

Let the concentrations be at the equilibrium values $x = x^0$, $y = y^0$ for some time and then let them be disturbed to small values

$$\xi \ll x, \quad \eta \ll y.$$

The second differential equation of the set (1) now gives

$$\frac{dy}{dt} = k_2\xi - b_2x^0 \simeq - b_2x^0$$

initially, while y starts at a value much smaller than b_2x^0. So y becomes negative.

5d A Single Goodwin Equation with Lag

In connection with a model to be discussed in Chapter 6 it is necessary to examine the special case in which all the linear equations in the set (1) relate to lag, namely

$$\frac{dx_1}{dt} = \frac{k}{1+x_{p+2}^{\rho}} - bx_1, \qquad (10)$$

$$\frac{dx_r}{dt} = a(x_{r-1}-x_r), \quad r = 2 \text{ to } p+2.$$

In particular it will be useful to obtain results indicating the value of ρ likely to give instability of the equilibrium point for particular p, and an estimate of the period as the instability sets in. These two requirements are equivalent to finding explicit results for a_c and ω_o in terms of the parameters appearing in (10). It becomes much easier to get ω_o in the special case b = a. Then the stability equation is

$$(\lambda+a)^{p+2} + \beta k a^{p+1} = 0.$$

Following Hunding (1974) this has roots

$$\lambda_m = -a + a\left[\frac{k\beta}{a}\right]^{1/p+2} . \exp\left(\frac{\pi i}{p+2} + \frac{2\pi i m}{p+2}\right), \quad m = 0 \text{ to } p+1,$$

and the two that become pure imaginary at the onset of instability are

$$\lambda_o, \lambda_{p+1} = -a+a\left[\frac{k\beta}{a}\right]^{1/p+2} . \exp\left(\pm\frac{\pi i}{p+2}\right) .$$

The notation used here is that

$$\beta = \frac{\rho x^{\rho-1}}{(1+x^{\rho})^2}, \quad x = x_{p+2}^0 ,$$

so that

$$\frac{\beta k}{a} = \rho\,\frac{(c-1)}{c}, \quad c = 1+x^{\rho} = k/ax.$$

Hence the critical value of ρ is

$$\rho = \left(\frac{c}{c-1}\right) \sec^{p+2}\left(\frac{\pi}{p+2}\right) . \qquad (11)$$

[For fixed ρ the critical value of a then depends on p.] The pure imaginary roots are

$$\pm i\omega_o = \pm ia \tan\left(\frac{\pi}{p+2}\right).$$

Hence the ratio $T_{osc}/\bar{T}$ is

$$\frac{2\pi}{p+1} \cot\left(\frac{\pi}{p+2}\right). \tag{12}$$

This is insensitive to p, being $2\pi/3$ for $p = 2$, and falling only to 2 as p becomes very large. [These values are very different from the values of this ratio in the logistic model.] Dropping the restriction to $b = a$, one can still find the critical value for ρ for small p values by using the Routh-Hurwitz determinants, derived from the stability equation

$$(\lambda+a)^{p+1}(\lambda+b) + \beta k a^{p+1} = 0.$$

With $p = 1$ this yields

$$\rho > \left(\frac{c}{c-1}\right)\frac{2(a+b)^2}{ab}, \tag{13}$$

and with $p = 2$

$$\rho > \left(\frac{c}{c-1}\right)\frac{8(a+b)^3}{b(3a+b)^2}. \tag{14}$$

5e Discrete Lag in the Goodwin Model

In this section I shall briefly summarise some results of an der Heiden (1978), who has examined the Goodwin equations with a discrete lag in each off-diagonal term,

$$\begin{aligned}\frac{dx_1}{dt} &= f(x_n(t-T_n)) - b_1x_1(t),\\ \frac{dx_i}{dt} &= k_ix_{i-1}(t-T_{i-1}) - b_ix_i(t),\ i = 2, \ldots n.\end{aligned} \tag{15}$$

He takes the function f(x) to be non-negative, continuous and monotone decreasing. He points out that the equations (15) reduce to a set with a single lag in the non-linear term, by the substitution

$$y_r(t) = x_r\left(t + \sum_{s=1}^{r-1} T_s\right),$$

and indeed simplify further by the substitution

$$v_r = a_r y_r$$

$$a_n = 1, \quad a_r = \prod_{s=r+1}^{n} k_r, \quad r = 1, \ldots n-1,$$

the end result being

$$\frac{dv_1}{dt} = a_1 f(v_n(t-T)) - b_1 v_1(t),$$

$$\frac{dv_i}{dt} = v_{i-1}(t) - b_i v_i(t) \qquad i = 2, \ldots n.$$

The stability equation is then

$$\prod_{i=1}^{n} (b_i+\lambda) + \alpha e^{-\lambda T} = 0, \tag{16}$$

where

$$\alpha = -a_1 f'(v_n^o) > 0.$$

an der Heiden shows that (16) has a root with positive real part, for the case n = 1, if and only if

$$\alpha^2 T^2 > \gamma^2 + T^2 b_1^2,$$

where γ is the solution of

$$\cot\gamma = -Tb_1/\gamma,$$

in the range $\frac{\pi}{2} < \gamma < \pi$. He shows that in the case n = 2 the corresponding requirement is that

$$\alpha^2 T^2 > \delta^4 + (b_1^2+b_2^2)\, T^2\delta^2 + T^4 b_1^2 b_2^2,$$

where δ is the solution of

$$\cot\delta = \frac{\delta^2 - b_1 b_2 T^2}{T(b_1+b_2)\delta},$$

in the range $0 < \delta < \pi$. Finally he shows that whenever there is such a root in these two cases (n = 1, n = 2) there is at least one non-constant periodic solution of the equations (15).

MODELS OF HAEMOPOIESIS

6a Wheldon's Model of Chronic Granulocytic Leukemia

The bone marrow contains stem cells which proliferate and also differentiate into precursors of the various types of blood cell, such as the erythrocytes (red cells) the granulocytes, and the monocytes (types of white cell). The maturing time in the marrow is for example about ten days for granulocytes and about three days for monocytes. In the case of monocytes further development takes place. These cells pass through the blood stream into tissue in about one day, and turn into macrophages. The number of cells of each type that are produced in the marrow of a healthy individual is adapted to the requirements of the body, red cell production being related to the need of tissues for oxygen, and white cell production to the need for defense against infection. General surveys of the topic of blood cell production, or haemopoiesis, can be found in the books by Metcalf and Moore (1971), Cline (1975) and Wickramasinghe (1975).

Control mechanisms must be present to relate the rate of production to the number of cells required. Hormonal factors are known to which the early precursors of the various cell types are sensitive, erythropoietin for precursors of erythrocytes and colony stimulating factor (CSF) for precursors of granulocytes and macrophages. The red cell case is the most fully understood. Oxygen deficiency triggers the production of erythropoietin in the kidney, and this substance circulates to the marrow and stimulates the production of red cells. In the case of granulocytes the situation is somewhat more obscure. It is known that monocytes in the blood stream, and macrophages in tissue, generate CSF. However there is also evidence that mature granulocytes in marrow secrete a substance that inhibits the production of granulocyte precursors. Thus both self-inhibition and cross-activation may be involved.

I shall look in this chapter at two models of blood cell production set up in terms of equations for the concentrations of mature cells in the marrow. As emphasised in 1c, this implies that the maturing times appear as lags. Both these models are concerned with diseases in which the concentration of granulocytes in the peripheral blood fluctuates in a roughly periodic manner. Apart from their possible relevance to therapeutic policies for these conditions, the main interest of the models is that breakdown of control mechanisms may give clues as to their normal functioning. In our context the particular interest is in exploring once more the relation between lags and periodicity, and the consequences of distributed as opposed to discrete lags. For, as in the context mentioned in 1f, one can not take maturing time as a fixed parameter for the whole population of one type of blood cell. Each individual granulocyte in the pool of mature cells in the marrow has a particular history and a well-defined maturing time, but two of them may differ for example in arising from a shorter and a longer sequence of cell splittings. Even if the same number of cell splittings have occurred, one must recognise that the cell cycles are not well synchronised throughout the population of proliferating and differentiating cells. The self-inhibiting effect may act in several successive cell cycles of the proliferating precursors, rather than on entry to the first one, giving an even stronger reason for using a broadly distributed delay.

The two diseases to which these models are applied are cyclical neutropenia and chronic granulocytic leukemia. In the first the concentration of granulocytes in peripheral blood falls, about every 20 to 25 days, to values much lower than normal. This means that the patient is very vulnerable to bacterial infections at these times. There is an animal model of the disease; collie dogs can have a genetically determined condition in which such dips in granulocyte concentration take place at intervals of about 11 days. Chronic granulocytic leukemia involves the overproduction of granulocytes. In some cases, against a rising background, peaks and dips have been observed with a period of 60 to 70 days.

Wheldon et al. (1974) relate the striking difference in the period in these two conditions to a change in the maturing time of the cells. They take the point of view that the oscillations in cyclical neutropenia represent transients in the response of the stable production process to some kind of recurring stress, the period being related to the lag that comes from the normal maturing time. They make the hypothesis that chronic granulocytic leukemia involves a drastic increase of the maturing time, by a factor of 3. This could have three effects.

I For some of this time cells are proliferating, so that an excessive amount of granulocytes are produced. While an essential part of the general argument, this is not put explicitly into the equations given below.

II Increasing the lag destabilises the steady state.

III Multiplying the lag by 3 multiplies the period by 3.

This model must stand or fall on the test of direct evidence that the leukemic cells, at any rate in this particular leukemia, have a drastically increased maturing time. Evidence on the development of these cells is hard to come by, for ethical rather than technical reasons, and there appears to be no suitable animal model of chronic granulocytic leukemia.

The model assumes a constant concentration of stem cells, and an inhibiting effect of the granulocytes in marrow on their own production, with a delay taken to be 7 days. There is also an inhibition, without delay, exercised by cells in the blood stream on the transfer of fresh cells from the marrow to the blood stream. This is thought to be necessary because of the relatively rapid response of the granulocyte concentration to the presence of infection. The two inhibition terms are taken in a form, reminiscent of the Goodwin model of Chapter 5, and the equations are

$$\frac{dx}{dt} = \frac{\alpha}{1+\beta x^{\gamma}(t-T)} - \frac{\lambda x}{1+\mu y^{\delta}}, \qquad (1)$$

$$\frac{dy}{dt} = \frac{\lambda x}{1+\mu y^{\delta}} - \omega y. \qquad (2)$$

Here x(t) refers to cells in marrow, and y(t) to cells in the blood stream. The model is rather rich in parameters, although these can be pinned down to some extent by our knowledge of the steady state values of x and y, the transit rate of cells from marrow to blood stream, and their half-life from entry to the blood stream until they leave it for tissue or die. The values assigned to the parameters are given in the table. The results obtained are that for T = 7 days T_{osc} = 20 days, while for T = 20 days T_{osc} = 60 days. In the first case the oscillations are damped, in the second case they grow.

Parameters adopted by Wheldon et al. in
equations (1) and (2)

α	1.1×10^{10} cells/kg/day	ω	2.43/day
β	10^{-12} (cells/kg)$^{-5/4}$	T	7 days, 20 days
λ	10 / day	γ	5/4
μ	4×10^{-8} (cells/day)$^{-1}$	δ	1

[cell concentrations are conventionally given in numbers per kg body weight]

Little is known about the spread of maturing times about the mean value, but it is of some interest to supplement these results with some obtained on the assumption of a broad spread. I have retained all the parameter values of the table, with T now interpreted as the mean value, $\bar{T} = 3/a$, for memory function $G_a^2(u)$. Going directly to numerical computation with the linear chain equations, I find that the ratio $T_{osc}/\bar{T}$ is about the same as the ratio T_{osc}/T found by Wheldon, but that the oscillations are damped even with $\bar{T}$ raised to 30 days.

These results are not to be regarded as a serious criticism of this model. One could probably modify parameters by acceptable amounts to bring the steady state normal condition nearer to instability, so as to counteract the relatively stabilising effects of distributed lag. Closer examination of the information available on cyclical neutropenia suggests that a more detailed model of that condition will be of some value.

6b <u>Two-lag Models of Cyclical Neutropenia</u>

In at least six cases in the literature [Page and Good (1957), Coutel et al. (1963), Videback (1965), Moore et al. (1974), Guerry et al. (1973), Meuret and Fliedner (1974)] it is clear that more than one type of blood cell concentration oscillates in cyclical neutropenia. In fact monocyte concentrations oscillate widely, roughly π out of phase with granulocyte concentrations. Moore et al. (1974) suggest a qualitative model to account for this, but do not formulate it mathematically. Here I shall present two alternative explicit versions of a model of the type they suggest, in which there is competition between the production of precursors of the two cell types. Any activating, or inhibiting, effect on granulocyte precursors is accompanied by a complementary inhibiting, or activating, effect on monocyte precursors. Two distinct lags appear in the model, because the maturing times for the two cell types are quite different. The evidence for a short maturing time for monocytes is rather recent. It comes from tracer experiments on a single individual [Meuret et al. (1974a)] and from the observation of low amplitude oscillations of the monocyte concentration, with period around 5 days, in 5 normal individuals [Meuret et al. (1974b)]. Investigation of the lags, and especially of the spread of lags, together with attempts to fit the major features of cyclical neutropenia, may elucidate some of the control mechanisms.

The features that I attempt to incorporate in the models are

I There should be a critical parameter which controls a change from stable equilibrium, taken to represent normal health, to unstable, taken to represent cyclical neutropenia.

II In the "unstable" regime, there should be a stable periodic solution, with both the concentrations oscillating.

III The period should be of the order of 20 to 25 days, for acceptable values of mean lags.

IV The two concentrations should be out of phase by about π.

V The granulocyte concentration should vary from a maximum near the normal value to a minimum reduced by at least a factor 10.

VI The monocyte concentration should vary from a minimum near the normal value (which is down by a factor of ten on the normal granulocyte concentration) to a maximum raised by about a factor 10.

VII The steady state values of the two concentrations, and of the transfer rates from marrow, should be realistic.

It is known that precursor peaks precede peaks in the blood stream cells, so I shall use the concentrations of mature cells in marrow as the only variables, regarding the control of marrow to blood stream transit as immaterial. Clearly the lags will be essential so far as II, III are concerned, while the assumed competition is intended to take care of the complementary character of V and VI. As indicated at the beginning of this chapter, there is evidence for two alternative modes of control of granulocyte production, and two models are set out that each use one of these.

Model A Mature granulocytes are assumed to inhibit production of granulocyte precursors. The cells from which these develop are committed either to the

FIG 9

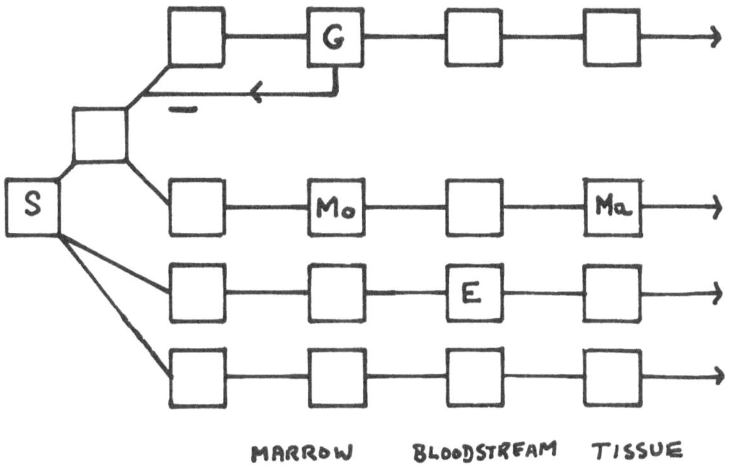

granulocyte path or to the monocyte path but cannot enter any other. So there is an effective activation of the monocyte precursor production. The model is schematically illustrated in Figure 9. A specific form of inhibition function is chosen to exploit what has been learned in Chapter 5. Taking x(t) as the monocyte concentration and y(t) as the granulocyte concentration, the equations are

$$\frac{dx}{dt} = \frac{\alpha\phi z^{\rho}}{1+\phi z^{\rho}} - \beta x, \tag{3}$$

$$\frac{dy}{dt} = \frac{\gamma}{1+\phi w^{\rho}} - \delta y. \tag{4}$$

As usual the delayed forms of y, which are z and w, have the same equilibrium value y^{o} as y. The equilibrium values x^{o} and y^{o} satisfy

$$\frac{\beta x^o}{\alpha} + \frac{\delta y^o}{\gamma} = 1, \tag{5}$$

an equation which brings out the "competitive" aspect of the model. The delayed form z is defined in terms of a distributed lag with memory function $G_a^p(u)$, for which the mean lag $(p+1)/a$ is the appropriate maturing time for monocytes, namely 3 days. The other delayed form w is defined in terms of a distributed lag with memory function $G_b^r(u)$, for which the mean lag $(r+1)/b$ is the maturing time for granulocytes, namely 10 days.

It is assumed that in normal health the inhibition function is a slowly varying function of the granulocyte concentration. In fact ρ is set equal to 1, so that there is no question of instability however long the lag. The transition to cyclical neutropenia is assumed to come from a sharpening of the dependence of this function on the granulocyte concentration, ρ becoming large. Then equation (4) has a periodic solution for sufficiently long lag. Clearly this equation can be solved independently of equation (3), which is a linear equation for $x(t)$ with a periodic forcing term provided by a function of the solution of (4). This term is delayed, and is averaged because of the spread of monocyte maturing times. The parameters α, β, γ, δ are fixed from the steady state values, employing $\rho = 1$, while ϕ is chosen so that instability is accompanied by falling mean value of $y(t)$. Before proceeding to numerical work, the results of 5d can be employed to estimate how large ρ must be to get instability, and what the period of the solution is likely to be. As expected from these results, the period is essentially determined by the mean lag, and is almost independent of the spread in the distributed lag. Numerical calculations are performed using the linear chain equations, either with $\rho = 10$ or with a step function, which amounts to taking ρ very large.

Some typical results are shown in Figure 10. Features V and VI are given by this model, but the phase relation of the two oscillations is not given very well. In fact to get the phase correct it seems to be necessary to set the two lags equal, which is unrealistic.

Model B The alternative model has monocytes activating the production of granulocytes and, by the assumption of competition, inhibiting their own production. The equations (3) and (4) are again used, but with $x(t)$ now referring to granulocyte concentration and y(t) to monocyte concentration. The mean lag in equation (4) must again be about 10 days to give a reasonable period. This can only be done by making an arbitrary, but in principle testable, assumption that on the average monocytes or macrophages in tissue have passed 7 days out of the marrow at the time at which they secrete the actuating factor.

Two points must be noted about this assumption. This extra delay must have a substantial spread, since there is nothing so far as we know crucial about the lapse of 7 days. There is in fact some evidence [Moore et al. (1974)] that already in their brief passage through the blood stream monocytes are a significant source of CSF. The delay in equation (3) will be a convoluted delay of the maturing time of

FIG10

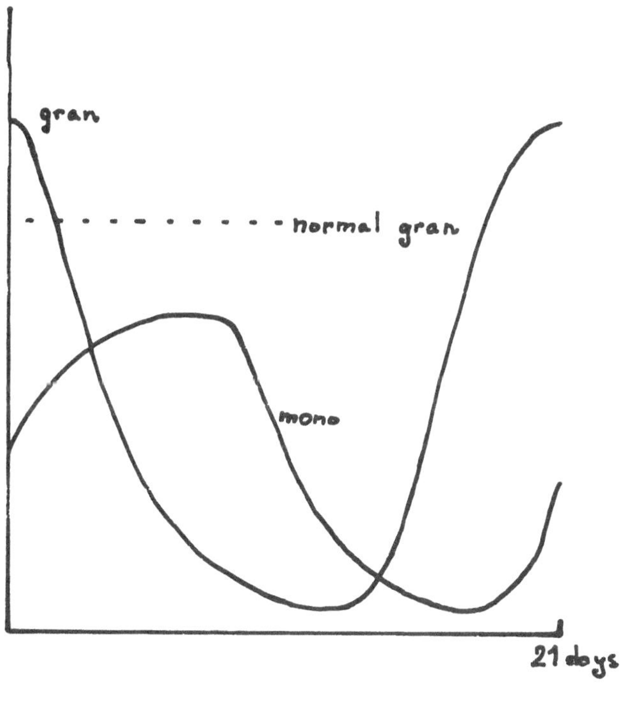

the granulocytes, with mean lag 10 days, and of the new delay of the monocytes/ macrophages in tissue, with mean lag 7 days. Using the memory functions $G_a^p(u)$ these can be combined into a single distributed lag of higher order, with total mean lag 17 days. It is significant that this is comparable with the period of the forcing term in equation (3). The averaging effect in the forcing term of equation (4) is now rather important, damping down the amplitude of the granulocyte oscillations considerably.

The assignment of parameters proceeds as in Model A, with ϕ now chosen so that instability is accompanied by falling mean value of x(t). Typical results are shown in Figure 11. It is clear that the feature V is not well reproduced. To

FIG 11

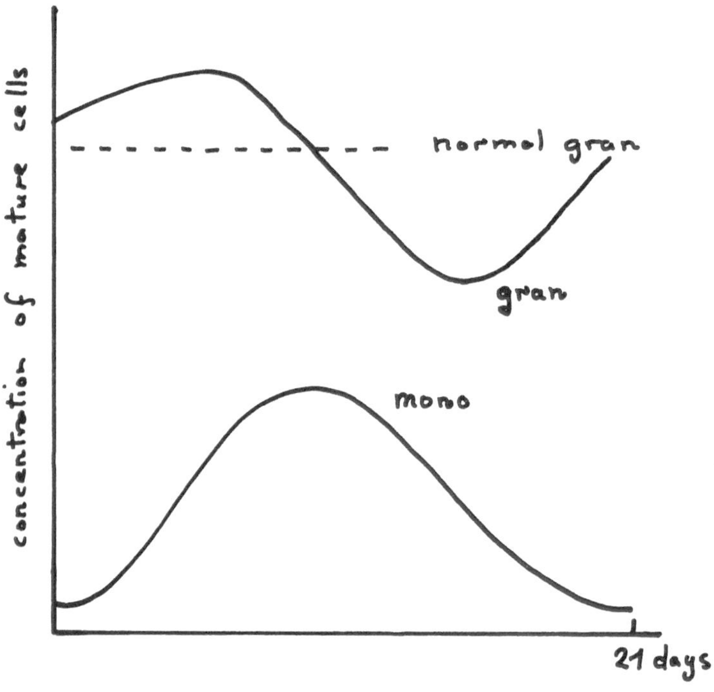

sum up, data on cyclical neutropenia favour Model A over Model B. The period is dependent only on the mean lag in the inhibition equation, and is not sensitive to the form of the memory function. The amplitude of the oscillations obtained from the activation equation is sensitive to the spread of lags because the effect of the averaging process depends on the ratio of the spread to the period. With memory function $G_a^p(u)$ at low p the spread is comparable to the mean lag, and the ratio of this to the period is 17/21 in Model B but only 1/7 in Model A. Both models have the defect of not giving the observed monocyte to granulocyte phase shift. Some further discussion of these models is presented in MacDonald (1978).

The possibility should be borne in mind that the monocyte peaks may be a secondary effect of infections setting in as the granulocyte concentration falls below normal. This would make Model B meaningless, and reduce Model A to a slightly more sophisticated variant of the models dealing with the granulocyte oscillations alone on the basis of self-inhibition. It should also be noted that the most detailed study of cyclical neutropenia so far reported indicates the presence of oscillations in other cell lineages [Guerry et al. 1973, Hoffmann et al. 1974].

6c Time Lag with Attrition; a Model of Cyclical Pancytopenia

I have emphasised in section 1e that models with lags, constructed as extensions of instantaneous models, are expected to leave equilibrium points unmoved, even if their stability character is changed. However there are exceptions to this, if the lag is accompanied by loss from the system, so that for example one replaces x(t), not by x(t-T) but by x(t-T)exp(-CT). The lag now not only changes the nature of the equations, from ordinary differential to delay-differential ones, but alters the value of at least one parameter, so that one may expect the equilibrium points to move. This has been called "attrition" in a general discussion of the consequences of such changes. [van Sickle (1977)]

The present discussion of models of cyclical haemopoiesis affords an opportunity for a brief discussion of a model with attrition. This was proposed by Mackey (1978) as a model of cyclical neutropenia, or rather of cyclical pancytopenia, since it requires all the populations of blood cells to oscillate about lowered mean values. He assumes that the reason for instability and oscillations lies in the death of cells in the proliferating pluripotent stem cell population. The lag time is the duration of the cell cycle, and thus is of the order of 1 day. The observed period of oscillation, which is much longer, is dependent in quite a complicated manner on a number of time constants including the lag.

Following on an established tradition in cell kinetics [Burns and Tannock (1970), Smith and Martin (1974)] Mackey takes two cell populations, with relationships indicated schematically in Figure 12. P(t) refers to cells that proceed through the sequence of events that make up the cell cycle, such as DNA synthesis and mitosis. N(t) refers to cells that are in a "resting" phase, with random exit back into the cycle, or into differentiated forms. All cells that survive the full cycle time T leave the cycling population for the resting one, but there is random loss from the cycling population. These assumptions lead to equations of the form

$$\frac{dN}{dt} = -DN - B(N)N + 2B(N(t-T))N(t-T)\exp(-CT),$$
$$\frac{dP}{dt} = -CP + B(N)N - B(N(t-T))N(t-T)\exp(-CT). \qquad (6)$$

Mackey takes the controlling function B in the form

$$B(N) = \frac{B_o\theta^n}{\theta^n + N^n}$$

FIG 12

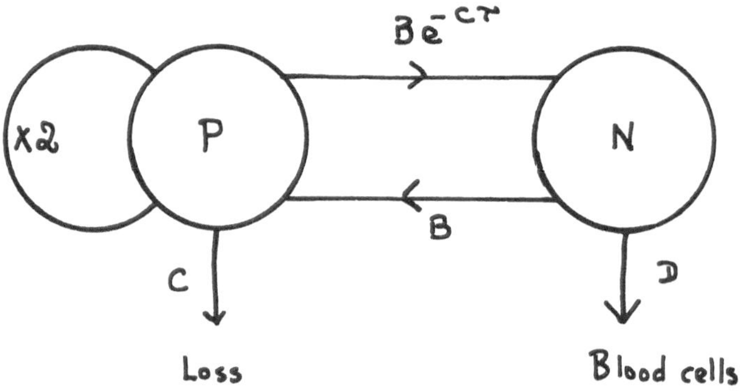

The equilibrium value of N is given by

$$N_o^n(C) = \theta^n \left[\frac{B_o}{D}(2e^{-CT}-1) - 1\right], \qquad (7)$$

which rises to

$$N_o^n(0) = \theta^n \left[\frac{B_o}{D} - 1\right]$$

as $C \to 0$. The equilibrium value of P is given by

$$P_o(C) = N_o(C)\,\frac{D}{C}\left\{\frac{1-e^{-CT}}{2e^{-CT}-1}\right\}, \qquad (8)$$

which rises to

$$P_o(0) = N_o(0)DT$$

as $C \to 0$.

Mackey can obtain oscillations for C = 0, for example by taking sufficiently high n. However he argues that the most realistic version of the model is one in which the primary cause of oscillations is loss of cells from the cycling population. As C is raised the following sequence of changes occurs in the qualitative behaviour of the solutions of equations (6). At first N_0 and P_0 fall slightly, but the equilibrium point remains stable. Then there are oscillations around an unstable equilibrium point. Finally the equilibrium point, now having $N_0(C)$ and $P_0(C)$ considerably below $N_0(0)$ and $P_0(0)$, is again stable. In pancytopenia induced by the drug cyclophosphamide in dogs [Morley and Stohlman 1970] this sequence of mild chronic pancytopenia, cyclical pancytopenia, and pronounced chronic pancytopenia is observed as dosage is increased.

An obvious extension of Mackey's model is to allow for the spread of cell cycle times in the population P(t) by taking a distributed lag. The stability aspects of this have been investigated by Rotenberg (1978) using the Laplace transform (31) of 2d.

A possible method can be given for interpreting observations of oscillating populations of several, but not all, cell types. This involves resonance between a low amplitude oscillation of stem cell origin, and damped oscillations of the different cell lineages. I have obtained some results in a crude model of this kind. In equation (4) the oscillating output DN(t) of the Mackey model (6) replaces the constant γ, while ρ is given the value 4, too small to give autonomous sustained oscillations. There are driven oscillations in the blood cell populations with amplitude and phase dependent on the difference between the natural frequency and the stem cell driving frequency. Extensive calculations of this general type are being carried out [personal communication from M.C. Mackey].

Knowledge of the control mechanisms of haemopoiesis is continually being extended [See for example Kurland et al. (1978)]. None of the models discussed in this Chapter may survive very long. However so long as action on early precursors by the secretions of mature cells remains a feature of these processes, time-lag models will offer a convenient shortcut in assessing the behaviour of this system.

PREDATION MODELS OF THE VOLTERRA TYPE

7 A great deal of attention has been paid, in the literature of mathematical biology, to periodicity in populations of prey and predator species, arising from the dynamics of their interaction rather than from external periodic forcing. Volterra's early work on this topic, together with his parallel work on competition, undoubtedly had a very stimulating effect on the development of quantitative ecology, and in particular of laboratory studies of two-species systems. Whether the continued preoccupation with this question has been really fruitful is debatable [Hutchinson (1975)].

Be that as it may, the results of Chapter 4 indicate that time lags in the dynamics of predation are liable to bear on periodicity. If a model of a single species population, which in its instantaneous form can only yield monotone increase towards the carrying capacity, acquires periodicity by means of a lag, then it is all the more likely that lag will have some effect on two-species oscillations. This aspect of the dynamics has received its due share of attention [Wangersky and Cunningham (1957), Caswell (1972), Ross (1972), May (1973, 1974), Bownds and Cushing (1975), MacDonald (1975), Cushing (1976), Knolle (1976), MacDonald (1977d), Arditi et al. (1977), Brauer (1977), Leung (1977), Cushing (1978), Morris (1978)]. Here I shall look briefly at the effect of two different lag terms. One is in the quadratic limiting term in the prey equation, and is directly comparable to that in the logistic model for one species. The other is in the interaction term in the predator equation, as suggested by the argument put forward in 1b, and as originally proposed by Volterra (1927).

The reader will hardly need reminding that Volterra's predation model, in its first instantaneous version,

$$\frac{dx}{dt} = \varepsilon x - \alpha xy, \qquad\qquad (1)$$
$$\frac{dy}{dt} = -\gamma y + \beta xy,$$

led to nested closed periodic orbits in the x, y plane, with peaks in the prey population x(t) leading peaks in the predator population y(t) by $\pi/2$. When a quadratic limiting term was introduced in the prey equation, so that it became

$$\frac{dx}{dt} = \varepsilon x(1-x/K) - \alpha xy, \tag{2}$$

Volterra found that however large K the trajectories spiral in to the point

$$x^0 = \gamma/\beta, \quad y^0 = (\varepsilon/\alpha)(1-\gamma/\beta K), \tag{3}$$

and there is no question of sustained oscillations. May (1973) pointed out that distributed lag in the quadratic term could restore oscillations, now of limit cycle type and so more biologically respectable than those in the original model. Morris (1978) has recently applied the Poincaré-Lindstedt method to this problem, with discrete lag. He obtains an approximation to the periodic solution in the form of an ellipse in the x,y plane

$$(x-x^o)^2 + (y-y^o)^2 \left(\frac{\omega_o\varepsilon}{y^o\beta}\right)^2 = B^2K^2. \tag{4}$$

Here the amplitude B is determined in a manner analogous to that of section 4e. The lowest order estimate of the period is $2\pi/\omega_o$, for

$$\omega_o = \tfrac{1}{2}\{x^o/K + \sqrt{[(x^{o2}/K^2) + 4(\alpha\beta/\varepsilon)\, x^o y^o]}\}, \tag{5}$$

and the critical delay is $\pi/2\omega_o$. This gives a rough estimate of the ratio T_{osc}/T of 4, as for the logistic model.

Returning to equations (1) Volterra (1927) introduced distributed lag in the predator equation, writing

$$\frac{dy}{dt} = -\gamma y + \beta yz, \tag{6}$$

with

$$z = \int_{-\infty}^{t} y(\tau)\, G(t-\tau)d\tau.$$

He gave no results for specific forms of memory function G, but was able to prove for general positive G the results that x(t) and y(t) fluctuate continually about mean values equal to the equilibrium values α/β, ε/α.

I have examined [MacDonald (1975)] this model with the special form of memory function $G_a^p(u)$, with and without the logistic term in the prey equation. Without the logistic term, and taking the simplest case p = 0, the equilibrium point is unstable however large a. This reflects the structural instability of the original Volterra equations (1). The trajectories spiral outwards with no sign of a limit cycle. When the logistic term is included, one has a balance between a stabilising effect (limit on prey numbers) and a destabilising effect (lag in prey-predator interaction) and a limit cycle ensues. This is found numerically for $p = 0$ and $p = 1$.

May (1974a) has shown that modified Volterra instantaneous equations with logistic growth of prey population (stabilising effect) and limitation of predator appetite (destabilising effect) give periodic behaviour. He uses a general argument, appropriate to a two-dimensional model, and equivalent to the criterion of Poincaré and Bendixson for the existence of a limit cycle. When lag is introduced into the prey-predator interaction term I find [MacDonald (1975)] that the parameter range allowing periodicity is increased, so the two destabilising effects cooperate. However in a more restricted parameter range a rather curious phenomenon is observed. As the mean lag increases from zero, the equilibrium point becomes unstable and a limit cycle of small amplitude is found. Further increase in mean lag causes the amplitude of this limit cycle first to increase and then to decrease. Finally even longer mean lag gives stable equilibrium once more.

Undoubtedly the most thorough exploration of lag effects in predation models is that described in the lecture notes of Cushing (1978). He works with distributed lag in both the interaction terms of the Volterra model, and includes limited prey population growth. He uses methods appropriate to general memory functions, supplemented by numerical work with $G_a^1(u)$. Thus the general consequences of lag in predation models are now well documented.

8a Stability Analysis

It is sometimes appropriate to set up a model in a form in which a new value of each variable, at time t, is defined as a function of the values of the variables at the earlier time t-T,

$$x_i(t) = h(x_1(t-T), \ldots x_n(t-T)). \tag{1}$$

This might be the appropriate form, for example, in which to model insect adult populations when there is one generation per year, separated by a phase of diapause. The complicated behaviour that can occur for solutions even of extremely simple equations of this type have only recently begun to be appreciated.

Equations of the type (1) may be regarded as extreme versions of equations with retarded variables. The stability analysis goes through in a similar manner to that presented in Chapter 2.

Again defining

$$\underset{\sim}{X} = \underset{\sim}{x} - \underset{\sim}{x}^o,$$

where $\underset{\sim}{x}^o$ is the equilibrium point, linearised equations for $\underset{\sim}{X}$ have the form

$$\underset{\sim}{X}(t) = B\underset{\sim}{X}(t-T). \tag{2}$$

An exponential solution is again appropriate, and requires the eigenvalue equation

$$R(\sigma) = |\bar{B} - I| = 0 \tag{3}$$

to hold. In this equation

$$\bar{B}_{ij} = B_{ij}\exp(-\sigma T).$$

[The eigenvalue is denoted by σ in this section, to reserve λ for use in equation (6) in agreement with the notation used by May.] The condition for a stable equilibrium point is that the roots σ of (3) have negative real part. Since (1) contains no differentiations to provide σ^r terms in (3), the function $R(\sigma)$ is a polynomial in $\exp(-\sigma T)$. Multiplying throughout by a sufficient number of factors $\exp(\sigma T)$ yields the polynomial equation

$$S(\mu) = 0, \tag{4}$$

in which $\mu = \exp(\sigma T)$. The condition for stability is now that all the roots μ of

(4) lie within the unit circle with centre at (0,0). $S(\mu)$ is of order n.

Methods for testing whether (4) satisfies this condition are given in the book by Porter (1967). There are many ways of mapping the inside of the unit circle onto the left hand half plane. A convenient one is by means of the transformation

$$\mu = \frac{\nu + 1}{\nu - 1}.$$

Applying this to (4) gives another polynomial equation,

$$T(\nu) = (\nu-1)^n S\left(\frac{\nu+1}{\nu-1}\right) = 0, \qquad (5)$$

and the Routh-Hurwitz tests can be applied to (5).

This Chapter is largely concerned with a model of a single insect species with adult population N, and generation span one year, such that to first approximation N obeys a simple difference equation

$$N(t) = \lambda N(t-1) \exp(-aN(t-1)), \quad \lambda > 1. \qquad (6)$$

The condition on λ is required so that a small population will grow. I shall take into account, as a second approximation, the phenomenon of extended diapause. In this a fraction α of the eggs laid each year lie two years rather than one before hatching, or there is an equivalent extended dormant phase at some later pre-adult stage. Equation (6) is then modified to the form

$$N(t) = (1-\alpha) \lambda N(t-1) \exp(-aN(t-1)) + \alpha\lambda N(t-2) \exp(-aN(t-2)). \qquad (7)$$

This may be regarded as introducing a lag, or as replacing a discrete lag by a distributed one, according to taste. The effects to be examined are stabilising ones, suggesting the second interpretation. On the other hand, (7) is of higher order than (6), and can be expressed as a pair of first order difference equations, which makes the change formally more like the introduction of a lag. The new pair of equations are

$$N(t) = (1-\alpha) \lambda N(t-1) \exp(-aN(t-1)) + M(t-1),$$

$$M(t) = \alpha\lambda N(t-1) \exp(-aN(t-1)).$$

The stability analysis of (6) and (7) follows. The equilibrium point of (6) is given by

$$N^o = \lambda N^o \exp(-aN^o),$$

so that

$$aN^{o} = \ln\lambda. \tag{8}$$

Letting $N(t) = N^{o} + n(t)$, $n(t)$ satisfies to linear terms the equation

$$n(t) = n(t-1)(1-aN^{o}) = n(t-1)(1-\ln\lambda).$$

The stability equation is simply

$$\mu = (1-\ln\lambda), \tag{9}$$

and μ lies inside the unit circle if $\lambda < e^2$. Recall that the model was formulated with the requirement that λ exceeds 1, so that it is not envisaged that μ should cross the unit circle at a point corresponding to $\lambda = 1$. Going to equation (7) the equilibrium point is

$$aN^{o} = aM^{o} = \ln\lambda.$$

Linearising around these values gives

$$n(t) = n(t-1)(1-\ln\lambda)(1-\alpha) + \alpha m(t-1),$$

$$m(t) = \alpha m(t-1)(1-\ln\lambda),$$

and the new stability equation is

$$\mu^2 + (1-\alpha)(\ln\lambda-1)\,\mu + \alpha(\ln\lambda-1) = 0. \tag{10}$$

Writing $\phi = \ln\lambda-1$, and applying the transformation from μ to ν, this gives a quadratic equation. The stability test is that all three terms have coefficients of the same sign. The coefficients are

$$1 + \phi = \ln\lambda > 0,$$

$$2(1 - \alpha\phi),$$

$$1 - \phi(1 - 2\alpha),$$

so the conditions for stability are

$$\ln\lambda < 1 + 1/\alpha,$$

$$\ln\lambda < \frac{2(1-\alpha)}{1-2\alpha}.$$

For a small fraction α undergoing extended diapause, the only significant condition is the second, or approximately

$$\ln\lambda < 2(1+\alpha). \tag{11}$$

8b <u>Conditions under which spreading the lag does not affect local stability</u>

Consider now the consequences of working with an equation formally identical with equation (6), but for which the condition $\lambda > 1$ is replaced by the condition

$\lambda < \Lambda$, where Λ is somewhere in the range 1 to e^2. In fact take the linearised version of (6) as

$$n(t) = n(t-1)(1-\ell n\lambda) = n(t-1)p,$$

where the emphasis is now on p around 1 instead of p around -1, so far as stability is concerned. The stability condition (9) is now interpreted as meaning $p < 1$. With the lag spread, the corresponding linearised equation is now

$$n(t) = (1-\alpha)\, pn(t-1) + p\alpha n(t-2),$$

and the stability equation is

$$\mu^2 - \mu p(1-\alpha) - p\alpha = 0.$$

Transforming to ν in the usual way gives

$$(\nu+1)^2 - (\nu^2-1)\, p(1-\alpha) - (\nu-1)^2\, p\alpha = 0.$$

The stability conditions are again that the coefficients of ν^2, ν and the constant term should all be positive, giving

$$\begin{aligned} 1 - p &> 0, \\ 1 + \alpha p &> 0, \\ 1 + p - 2\alpha p &> 0. \end{aligned} \qquad (12)$$

Taking α again in the range 0 to 1, these are all consequences of the stability condition for the original first order equation. Thus the effect of improving stability by spreading the lag is not universal.

I shall now present a general result for the conditions under which an extension from a "one step" difference to one with steps of one, two, . . gives no change in the stability conditions. Consider two models, which are both linear sets of equations, both of dimension n,

$$\underset{\sim}{x}(t) = A\, \underset{\sim}{x}(t-1), \qquad (13)$$

and

$$\underset{\sim}{x}(t) = \sum_{k=1}^{m} A_k\, \underset{\sim}{x}(t-k). \qquad (14)$$

The restriction is made that the net influence of all earlier times is the same in the two models, in the sense that

$$A = \sum_{k=1}^{m} A_k.$$

Then the stability of the equilibrium point (0, . . 0) in the case of equations (14) is implied by, and implies, the stability of this point in the case of equations (13), provided that

I The sign of A_k^{ij} is the same for all k and given i, j.

II A is a Morishima matrix.

A is said to be a Morishima matrix if one can find a permutation matrix P for which

$$A^+ = P'AP = \begin{bmatrix} A_{11}^+ & A_{12}^+ \\ A_{21}^+ & A_{22}^+ \end{bmatrix},$$

where A_{11}^+ is a q × q non-negative matrix, A_{22}^+ is a n-q × n-q non-negative matrix, and A_{12}^+ and A_{21}^+ are non-positive matrices. Thus Morishima matrices include as a special case positive matrices. Consequently this theorem covers as a trivial special case the stability analysis of (6) and (7) when λ is allowed to fall below 1, but not allowed to rise above e^2.

8c Chaos in Discrete Dynamical Systems

The more complicated kinds of behaviour possible in dynamical systems have attracted a great deal of attention recently. A frequently cited example is that of the Lorenz model [Lorenz (1963)], a much simplified set of equations for the motion of a fluid in a horizontal layer of uniform depth with fixed temperature difference between the two surfaces. There are three ordinary differential equations of which two are non-linear. In an appropriate range of parameters there are two stable equilibrium points. Each of these corresponds, in the original full set of partial differential equations, to smooth convective flow of the fluid, in opposite senses. A Hopf bifurcation occurs as these points become unstable, at a critical value of one parameter. As discussed by Marsden and McCracken (1976) Lorenz' results correspond to a case in which this bifurcation is sub-critical, and are in fact typical of the sub-critical bifurcation in this set of equations. Lorenz finds trajectories that perform several circuits around one equilibrium point, then jump to making circuits around the other. The number of circuits between these jumps, and the precise manner of flipping over, differ each time. Loosely speaking, the net effect is to fill up a region of the phase space. A detailed topological description of the Lorenz trajectories is quite complicated [Rand (1978)].

Many other examples are now known, involving sets of three or more differential equations, but no direct application to useful biological models has yet been made. Some potentially useful possibilities are outlined by Rössler (1977). In the biological context more attention has been paid to the analogous behaviour of the solutions to simple non-linear difference equations, of which the best known is equation (6) of this chapter. It should be noted that at least three first order differential equations are needed for this type of behaviour, because trajectories for such equations cannot cross except at equilibrium points. There is no such restriction for difference equations, or delay-differential equations.

In introducing this type of pseudo-random or chaotic behaviour to the field of population dynamics, May (1974b) was concerned, as Hutchinson (1948) had been, with the interpretation of fluctuations in the populations in terms of a model in which small populations will grow exponentially, because of the innate fecundity of individuals, while large ones will crash because the resources available are limited. While Hutchinson was concerned with the time-delayed logistic equation, May examined difference equations roughly analogous to it, such as equation (6). Writing one of these as

$$N(t) = f(N(t-1)), \tag{15}$$

the general features of solutions are similar for a variety of functions f satisfying

a) $f(N) > 0$ for $N > 0$, and $f(0) = 0$.

b) $\frac{df}{dN} = 0$ at one value of N.

c) There is one fixed point, $N^o = f(N^o)$.

d) f involves a parameter λ such that increasing λ changes the shape of the peak, giving in particular larger negative slope at the fixed point, as shown in Figure 13.

A detailed discussion of the requirements for bifurcations, of the type to be described, is given by Allwright (1978).

As the slope at N falls through -1, the equilibrium point becomes unstable. There are two new stable points of the equation

$$N(t) = g(N(t-2)) = f(f(N(t-2))), \tag{16}$$

and the solution of (15), starting from any initial positive value of N, tends towards a stable two-point oscillation. With continued increase of λ, this solution in turn becomes unstable, and is replaced by a stable four-point oscillation. Oscillations of higher and higher period, always a power of 2, replace each other at smaller and smaller increments in λ. Then a critical λ is reached, beyond which solutions of no

FIG 13

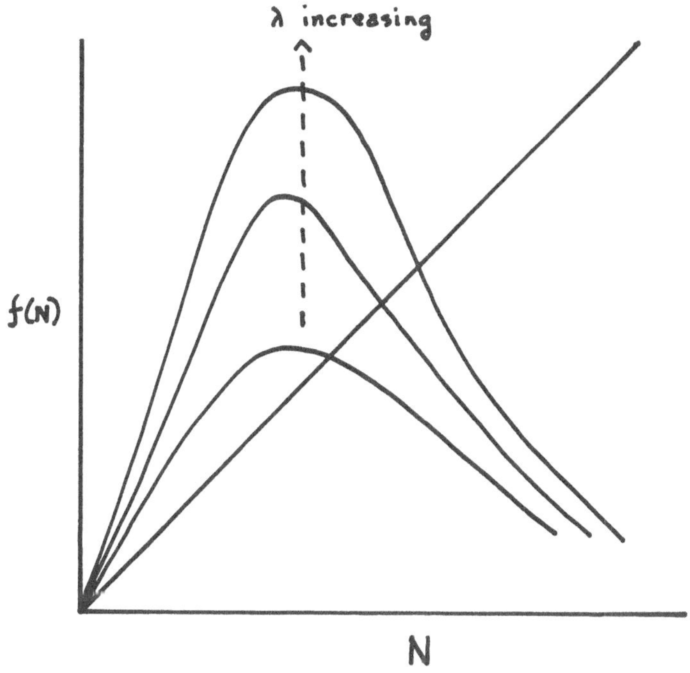

apparent periodicity can occur. At slightly higher λ, another qualitative change occurs. Solutions of any period can be found, as well as the periodic ones. Within any particular (and very narrow) "window" of λ values one solution will be stable and this will ultimately be attained by trajectories starting from essentially every initial value. The convergence to this solution is very slow and numerical results present the appearance of a sample of points from a continuous distribution. This stage is usually termed "chaos".

Similar or more elaborate behaviour has been found in a number of other models for one population or two populations, and the field has been reviewed by May (1976b).

8d Extended diapause in a single species population model

The results of the previous section imply that there are more ways than had been thought of interpreting fluctuations in a population. These may be due to

I demographic stochasticity, that is to the fact that numbers can only change by unit steps, and that birth and death rates are better thought of as probabilities rather than as fixed parameters.

II effects of random disturbances from the environment, that is to say from interactions not explicitly built into a deterministic model, possibly including interactions with other species as well as with the physical environment.

III purely deterministic dynamics functioning in the chaotic regime.

An attempt was made by Hassell et al. (1976) to assess whether III is a real possibility for insect populations. They fitted a two-parameter difference equation model, of which (6) is a special case, to 24 examples drawn from natural and laboratory populations. Only one example, the laboratory population studied by Nicholson (1954), yielded parameter values lying in the chaotic regime. Only one of the natural populations even lay in the regime of periodicity.

Another approach to this question is to ask what effects can be simply incorporated into the model that might be expected to reduce the regime of chaotic behaviour, for example in equation (6) to defer the onset of chaos to larger λ. Naturally effects are sought that are biologically plausible. It would seem that any effect that can act to damp out the effect of environmental disturbances might also stabilise the model against chaos. Since the onset of chaos is fore-shadowed by an increase of the maximum to minimum ratio in the periodic solutions, filling in minima and cutting down maxima may act to defer the onset of chaos.

As a specific example let us return to time lags and the discussion in 8a. It is not uncommon to find that an appreciable fraction of some pre-adult stage of an insect with normally annual generations lie dormant for a whole year, so that the corresponding fraction of the new adults emerge after two years rather than one. [Clark et al. (1967)] If the adult population is wiped out one year by environmental injury, the population will survive by virtue of the fraction undergoing this extended diapause. In a similar manner, facultative diapause is thought to be a regulating feature in populations of intestinal parasites [Schad (1977)].

Correspondingly, if the population runs into a sequence of very high and very low values, in a deterministic model with fixed diapause, then these might be expected to be less pronounced if the extended diapause is significant. This is the reason for investigating [MacDonald (1976)] whether equation (7), with α of the order 0.05 or

FIG 14

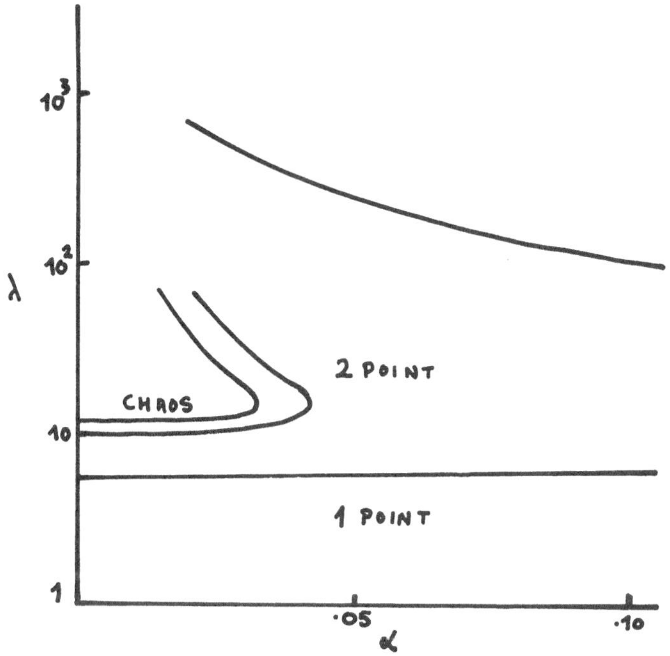

0.1, shows much difference from equation (6) with regard to stability of the equilibrium point, stability of simple periodic solutions, and λ value necessary for chaos.

The first question was examined in section a, and equation (11) shows that this is a rather small effect for such small values of α. To go beyond this analytically was not possible, but some interesting results were obtained by computation. These are displayed in Figure 14. The range of stability of the simplest periodic solutions is very considerably increased. As α falls to zero one recovers the results corresponding to equation (6), by way of a region in which solutions of higher period, and chaotic solutions, appear and then disappear as λ is raised. My conclusion is that if claims are to be made that actual insect populations fluctuate for reason III, the models used should be examined for their susceptibility to this stabilising effect, or others that may be appropriate. For example I find that a similar effect is produced by allowing mixing between locally breeding sub-populations, which is liable to be a feature of natural populations.

So far as the mixing of generations is concerned, its relevance to the concerns of this book is as an example, of a somewhat unusual kind, of the stabilising effect of distributed, as opposed to discrete lag. Recently some results have been presented that make it possible to illustrate the analogous stabilising effect in the case of equations of the delay-differential and integro-differential types.

8e Analogous Treatment of a Functional Differential Equation

It has been pointed out by Mackey and Glass (1977) that a delay-differential equation with a "hump", analogous to equation (6) can show a sequence of bifurcations comparable to those investigated by May (1974b). When the solutions are plotted as trajectories in a plane for which the coordinates are x(t) and x(t-T), it is found that as the hump becomes more pronounced a stable equilibrium point gives way to a simple closed trajectory, this in turn to a double loop, and so on. Eventually aperiodic trajectories are found. This is very different from the sudden jump into chaos of the model of Lorenz (1963). Rössler (1976) has presented a set of three ordinary differential equations of which the solutions exhibit a somewhat similar sequence of bifurcations.

The equation which Mackey and Glass study is

$$\frac{dx}{dt} = \frac{\lambda x(t-T)}{\alpha + x^n(t-T)} - \gamma x(t). \tag{17}$$

They present [Glass and Mackey (1978)] some results for $\alpha = \gamma = 1$, $\lambda = T = 2$, with n increasing to sharpen the hump. The equilibrium point, x = 1, is stable up to n = 5.04. The subsequent results are summarised in the Table, in which 3, for example, stands for a triple loop.

Results of Glass and Mackey (1978) on solutions of (17)

n	7	7.75	8.50	8.79	9.65	9.6575	9.76	10.0
trajectory	1	2	2	4	chaos	3	6	chaos

I have obtained a few results for the equation analogue to (17) with distributed lag, using the linear chain trick. I take a = 3/2 and p = 2, to give $\bar{T} = 2$, and $\lambda = 2$, $\alpha = \gamma = 1$. For this rather broadly distributed lag there is a marked expansion of the range of values of n over which the simplest periodic solution is found. From the table one sees that with discrete lag the full range from a stable equilibrium point to chaos is covered as n rises from 5 to 10. In my case the equilibrium point becomes unstable at n = 10.2, and up to n = 50 all the solutions computed are simple loops.

This method is not suitable for numerical investigation of what happens as discrete lag is approached. It has been shown [personal communication from P.E. Rapp] that a chain of equations resembling (1) of 5a, but with the monotonic non-linear function in the first equation replaced by a suitable non-monotonic function, can exhibit a strange attractor.

One can indeed formally set up equations having chaotic solutions and incorporating a zero order (p = 0) delay. The Lorenz equations, for example, in the usual notation are

$$\frac{dX}{dt} = \sigma(Y-X),$$

$$\frac{dY}{dt} = -XZ + rX - Y, \qquad (18)$$

$$\frac{dZ}{dt} = XY - bZ.$$

However in view of the original physical context in which X, Y and Z are defined it would be unnatural to interpret X as delayed Y. Curiously enough, Shimizu and Morioka (1978) have found it useful to convert (18), by a change of variables, into the form

$$\frac{d^2x}{dt^2} + (\sigma+1)\frac{dx}{dt} + \sigma(r-1)\ x(x^2-1+m),$$

$$\frac{dm}{dt} = (2\sigma-b)\ x^2-bm, \qquad (19)$$

which can be thought of as a linear damped oscillator with a force term incorporating delayed dependence on x.

Another recent development with potential applicability in biological models is the investigation by other Japanese workers [Tomita and Kai (1968), Fujisaka and Yamada (1978)] of chaotic behaviour of a forced non-linear oscillator and of sets of coupled non-linear oscillators. One can not exclude the possibility that forced or coupled "delay oscillators", as required by generalisations of the models of Chapter 6, may display chaotic behaviour.

Supplementary Bibliography

Apart from those papers cited in the text, it may be useful to list others which deal with lags in biological models. The titles are given to indicate the variety of contexts in which lags may be significant. List I contains papers that employ the linear chain trick, for low values of p. Some of these explicitly refer to memory functions G_a^0 and G_a^1, while others merely suggest an analogy between lag and the introduction of an extra stage in some process. List II contains papers that use other methods. Papers on epidemiology cited by Banks are not separately cited here.

I

Aizawa, Y. and Y. Kobatake (1976) Bull. Math. Biol. 38, 351. A theory of oscillatory instability in two-state systems in relation to nerve excitation.

an der Heiden, U. (1976) Math. Biosci. 31, 275. Stability properties of neural and cellular control systems.

Gomatam, J. and N. MacDonald (1975) Math. Biosci. 24, 247. Time delay and stability of two competing species.

MacDonald, N. (1976) Biotech. Bioeng. 18, 805. Time delays in simple chemostat models.

Rescigno, A. and C. de Lisi (1977) Bull. Math. Biol. 39, 487. Immune surveillance and neoplasia II A two-stage mathematical model.

Wörz-Busekros, A. (1978) S.I.A.M. J. Appl. Math. 35, 123. Global stability in ecological systems with continuous time delay.

II

Banks, H.T. (1975) Modelling and Control in the Biomedical Sciences. Lecture Notes in Biomathematics Vol. 6 Springer Verlag, Berlin.

Barclay, H. and P. van den Driesche (1975) J. Theor. Biol. 51, 347. Time lags in ecological systems.

Coleman, B.D. and G.H. Renninger (1974) Proc. Nat. Acad. Sci. 71, 2887. Theory of delayed lateral inhibition in the compound eye of Limulus.

Coleman, B.D. and G.H. Renninger (1975) J. Theor. Biol. 51, 243 and 267. Consequences of delayed lateral inhibition in the retina of Limulus, I and II.

Coleman, B.D. and Renninger, G.H. (1975) Istituto Lombardo di Scienze e Lettere, Rend. Cl. Sci. (A) 109, 91. Periodic solutions of a non-linear functional equation describing neural interactions.

Coleman, B.D. and G.H. Renninger (1976) J. Math. Biol. 3, 103. Theory of the Response of the Limulus retina to periodic excitation.

Cooke, K.L. and J.A. Yorke (1972) in Ordinary Differential Equations (ed. L. Weiss) Academic Press, New York. Equations modelling population growth, economic growth and gonorrhea epidemiology.

Dribov, B.F., M.A. Livshitz and M.V. Volkenstein (1977) J. Theor. Biol. 65, 609. Mathematical model of immune processes.

Gatica, J.A. and P. Waltman A singular functional equation arising in an immunological model. In Ordinary and Partial Differential Equations (ed. W.N. Everitt and B.D. Sleeman) Lecture Notes in Mathematics 564, Springer Verlag, Berlin.

Hadeler, K.P. (1976) J. Math. Biol. 3, 197. On the stability of the stationary state of a population growth equation with time lag.

Israelson, D. and A. Johnson (1967) Physiol. Plant. 20, 957. A theory for circumnutations in Helianthus annuus.

Israelson, D. and A. Johnson (1968) Physiol. Plant. 21, 282. Application of theory for circumnutations to geotrophic movement.

Johnson, A. and H.G. Karlson (1972) J. Theor. Biol. 36, 153. A feedback model for biological rhythms.

King-Smith, E.A. and A.A. Morley (1970) Blood 36, 254. Computer simulation of granulopoiesis; normal and impaired granulopoiesis.

Kirk, J., J.S. Orr and C.S. Hope (1968) Brit. J. Haem. 15, 35. A mathematical analysis of red blood cell and bone marrow stem cell control mechanisms.

Ladde, G.S. (1976) J. Theor. Biol. 61, 1. Stability of model ecosystems with time delay.

Lee, K.L. and E.R. Lewis (1976) I.E.E.E. Trans. Bio. Eng. 23, 225. Delay time models of population dynamics with applications to schistomiasis control.

London, W.P. and J.A. Yorke (1973) Amer. J. Epid. 98, 453 and 469. Recurrent epidemics of measles, chickenpox and mumps, I, II.

May, R.M., G.R. Conway, M.P. Hassell and T.R.E. Southwood (1974) J. An. Ecol. 43, 747. Time delays, density dependence and single species oscillations.

May, R.M. and R.M. Anderson (1978) J. An. Ecol. 47, 249. Regulation and stability of host-parasite population interactions II Destabilising processes.

Mazanov, A. (1973) Aust. J. Comp. 5, 74. Numerical method for solving population equations containing variable time lags.

Mazanov, A. (1976) J. Theor. Biol. 59, 429. Stability of multi-pool models with lags.

Moels, W.D. (1977) Math. Biosci. 34, 237. A mathematical model for the growth of cell cultures.

Nazarenkov, V.G. (1976) Biofisika 21, 352. Effect of delay on auto-oscillations in cell populations.

Smith, H.L. (1977) J. Math. Biol. 4, 69. Periodic solutions of a delay integral equation simulating epidemics.

Stein, R.B. and M.N. Oğuztöreli (1976) Biol. Cyber. 22, 147. Tremor and other oscillations in neuro-muscular systems.

Stirzaker, D.R. (1976) J. Inst. Math. App. 18, 135. On models for an epidemic.

Walther, H.O. (1976) J. Math. Biol. 1, 227. Existence of a non-constant periodic solution of a non-linear autonomous functional differential equation representing the growth of a single species population.

Waltman, P. (1974) Threshold Models in the Theory of Epidemics. Lecture Notes in Biomathematics Vol. 1. Springer Verlag, Berlin.

REFERENCES

Allwright, D.J. (1977a) Math. Proc. Camb. Phil. Soc. 82, 453.

Allwright, D.J. (1977b) J. Math. Biol. 4, 363.

Allwright, D.J. (1978) S.I.A.M.J. App. Math. 34, 687.

an der Heiden, U. (1978) Preprint, Department of Mathematics, Tübingen University.

Arditi, R., J.M. Abillon and J.V. da Silva (1977) Math. Biosci. 33, 107.

Beddington, J.R. and R.M. May (1975) Math. Biosci. 27, 109.

Bellman, R. and K.L. Cooke (1963) Differential-Difference Equations. Academic Press, New York.

Bownds, J.M. and J.M. Cushing (1975) Math. Biosci. 26, 41.

Brauer, F. (1977) Math. Biosci. 33, 345.

Burns, F.J. and I.F. Tannock (1970) Cell Tissue Kinet. 3, 321.

Caperon, J. (1969) Ecology 50, 188.

Caswell, H. (1972) J. Theor. Biol. 34, 419.

Clark, L.R., P.W. Geier, R.D. Hughes and R.F. Morris (1967) The Ecology of Insect Populations. Methuen, London.

Cline, M.J. (1975) The White Cell. Harvard Univ. Press, Cambridge, Mass.

Cooke, K.L. (1977) Preprint, Department of Mathematics, Pomona College.

Coutel, Y., H. Morel and G. Thomet (1963) N. Rev. Fr. Haem. 13, 719.

Cronin, J. (1977) S.I.A.M. Rev. 19, 100.

Cunningham, A. (1977) Thesis, Department of Applied Physics, Strathclyde University.

Cunningham, W.J. (1956) Proc. Nat. Acad. Sci. 42, 699.

Cushing, J.M. (1975) S.I.A.M.J. Math. Anal. 6, 433.

Cushing, J.M. (1976a) Math. Biosci. 31, 143.

Cushing, J.M. (1976b) S.I.A.M.J. App. Math. 31, 251.

Cushing, J.M. (1978) Integro-Differential Equations and Delay Models in Population Dynamics. Lecture Notes on Biomathematics, Volume 20. Springer Verlag, Berlin.

Dean, A.C.R., D.C. Ellwood, C.G.T. Evans and J. Melling (1976) Continuous Culture 6; Applications and New Fields. Society of Chemical Industry, London.

El'sgol'ts, L.E. and S.B. Norkin (1973) An Introduction to the Theory and Application of Differential Equations with Deviating Argument. Academic Press, New York.

Fargue, D. (1973) C.R. Acad. Sci. Paris B277, 471.

Finn, R.K. and R.F. Wilson (1953) J. Agric. Food Chem. 2, 66.

Forrester, J.W. (1961) Industrial Dynamics. M.I.T. Press, Cambridge, Mass.

Fujisaka, H. and T. Yamada (1978) Phys. Lett. A66, 450.

Gilbert, L.E. and P.H. Raven (1975) Co-evolution of Animals and Plants. Univ. of Texas Press, Austin.

Glass, L. and M.C. Mackey (1978) in Bifurcation Theory and Applications in Scientific Disciplines (ed. O. Gurel and O.E. Rössler) Ann. N.Y. Acad. Sci. in press.

Goel, N.S., S.C. Maitra and E.W. Montroll (1971) Revs. Mod. Phys. 43, 231.

Goel, N.S. and N. Richter-Dyn (1974) Stochastic Models in Biology. Academic Press, New York.

Goldbeter, A. and G. Nicolis (1976) Prog. Theor. Biol. 4, 66.

Goodwin, B.C. (1965) Adv. Enz. Reg. 3, 425.

Griffiths, J.S. (1968) J. Theor. Biol. 20, 202.

Guerry, D., D.C. Dale, M. Omine, S. Perry and S.M. Wolff (1973) J. Clin. Invest. 53, 3220.

Hale, J. (1977) Theory of Functional Differential Equations. Springer Verlag, Berlin.

Hassell, M.P., J.H. Lawton and R.M. May (1976) J. An. Ecol. 45, 471.

Hastings, S.P., J.J. Tyson and D. Webster (1977) J. Diff. Equ. 25, 39.

Hoffman, H.J., D. Guerry and D.C. Dale (1974) J. Interdisc. Cycle Res. 5, 1.

Holst, P.A. (1969) Simulation 12, 227.

Hsu, C.S. (1970) J. App. Mech. 37, 259.

Hunding, A. (1974) Biophys. Struct. Mechanism 1, 47.

Hutchinson, G.E. (1948) Ann. N.Y. Acad. Sci. 50, 221.

Hutchinson, G.E. (1975) in Ecology and Evolution of Communities (ed. M.L. Cody and J.M. Diamond) Harvard Univ. Press, Cambridge, Mass.

Jones, G.S. (1962) J. Math. Anal. App. 4, 440; 5, 435.

Kazarinoff, N.D., Y.H. Wan and P. van den Driesche (1978) J. Inst. Math. App. 21, 461.

Knolle, H. (1976) Math. Biosci. 31, 351.

Kostitzin, V.A. (1934) Symbiose, Parasitisme et Evolution. Hermann, Paris.

Kostitzin, V.A. (1937) C.R. Acad. Sci. Paris 204, 1683.

Kurland, J.I., R.S. Bockman, H.E. Broxmeyer and M.A.S. Moore (1978) Science 19, 552.

Landahl, H.D. (1969) Bull. Math. Biophys. 31, 775.

Langford, W.F. (1977) Numer. Math. 28, 171.

Leung, A. (1977) J. Diff. Equ. 26, 391.

Levins, R. (1975) in Ecology and Evolution of Communities (ed. M.L. Cody and J.M. Diamond) Harvard Univ. Press, Cambridge, Mass.

Lewis, E.R. (1972) Ecology 53, 797.

Lewis, E.R. (1977) Ecology 58, 738.

Lorenz, E.N. (1963) J. Atmos. Sci. 20, 130.

MacDonald, N. (1975) Math. Biosci. 28, 321.

MacDonald, N. (1976) Math. Biosci. 31, 255.

MacDonald, N. (1977a) Int. J. Systems Sci. 8, 467.

MacDonald, N. (1977b) J. Theor. Biol. 65, 727.

MacDonald, N. (1977c) J. Theor. Biol. 67, 549.

MacDonald, N. (1977d) Math. Biosci. 33, 227.

MacDonald, N. (1978) in Biomathematics and Cell Kinetics (ed. A.J. Valleron and P.D.M. Macdonald) North Holland, Amsterdam.

Mackey, M.C. (1978) Blood 51, 941.

Mackey, M.C. and L. Glass (1977) Science 197, 287.

Marsden, J.E. and M. McCracken (1976) The Hopf Bifurcation and its Applications. Springer Verlag, Berlin.

May, R.M. (1973) Ecology 54, 315.

May, R.M. (1974a) Stability and Complexity in Model Ecosystems. Princeton Univ. Press, Princeton.

May, R.M. (1974b) Science 186, 645.

May, R.M. (1976a) Theoretical Ecology, Blackwell, Oxford.

May, R.M. (1976b) Nature 261, 459.

Mazanov, A. and K.P. Tognetti (1974) J. Theor. Biol. 46, 271.

Mees, A.I. and P.E. Rapp (1978) J. Math. Biol. 5, 99.

Metcalf, D. and M.A.S. Moore (1971) Haemopoietic Cells. North Holland, Amsterdam.

Meuret, G., J. Bammert and G. Hoffmann (1974) Blood 44, 801.

Meuret, G., C. Bremer, J. Bammert and J. Ewen (1974) Cell Tissue Kinet. 7, 223.

Meuret, G. and T.M. Fliedner (1974) Blood 43, 565.

Michel, J.F. (1969) Adv. Parasitol. 7, 211.

Miller, R.K. (1972) in Delay and Functional Differential Equations (ed. K. Schmitt) Academic Press, New York.

Monod, J. (1950) Ann. Inst. Pasteur 79, 390.

Montroll, E.W. (1972) Proc. Nat. Acad. Sci. 69, 2532.

Moore, M.A.S., G. Spitzer, D. Metcalf and G. Penington (1974) Brit. J. Haem. 27, 47.

Morley, A.A. and F.S. Stohlman (1970) New Eng. J. Med. 282, 643.

Morris, H.C. (1976) J. Inst. Math. App. 18, 15.

Morris, H.C. (1978) Int. J. Systems Sci. in press.

Nicholson, A.J. (1954) Aust. J. Zoo. 2, 9.

Page, A.R. and R.A. Good (1957) Amer. J. Dis. Child. 94, 623.

Pontriagin, L.S. (1942) Izv. Akad. Nauk. S.S.S.R. Ser. Mat. 6, 115; also Amer. Math. Soc. Trans. 2cd. Ser. 1, 95.

Poore, A.B. (1976) Arch. Rat. Mech. Anal. 60, 371.

Porter, B. (1967) Stability Criteria for Linear Dynamical Systems. Oliver and Boyd, Edinburgh.

Rand, D. (1978) Math. Proc. Camb. Phil. Soc. 83, 451.

Rescigno, A. and I.W. Richardson (1973) in Foundations of Mathematical Biology (ed. R. Rosen) Academic Press, New York.

Ross, G.G. (1972) J. Theor. Biol. 37, 477.

Rössler, O.E. (1976) Z. Nat. 31a, 1664.

Rössler, O.E. (1977) Z. Nat. 32a, 607.

Rotenberg, M. (1978) Preprint, University of California, La Jolla.

Savageau, M.A. (1976) Biochemical Systems Analysis. Addison Wesley, Reading, Mass.

Schad, G.A. (1977) in Regulation of Parasite Populations (ed. G.W. Esch.) Academic Press, New York.

Shimizu, T. and N. Morioka (1978) Phys. Lett. 66A, 182.

Smith, J. and L. Martin (1974) Proc. Nat. Acad. Sci. 70, 1263.

Swinnerton-Dyer, Sir P. (1977) Math. Proc. Camb. Phil. Soc. 82, 469.

Tarr, D.G. (1976) Econometrica 44, 597.

Thingstad, T.F. and T.I. Langeland (1974) J. Theor. Biol. 48, 149.

Tomita, K. and T. Kai (1978) Phys. Lett. A66, 91.

Tyson, J.J. (1975) J. Math. Biol. 1, 311.

Tyson, J.J. and H.G. Othmer (1978) Prog. Theor. Biol. 5, 1.

van Kampen, N.G. (1976) Adv. Chem. Phys. 24, 245.

van Sickle, J. (1977) I.E.E.E. Trans. S.M.C. 7, 635.

Videbaek, A. (1965) Acta Med. Scand. 172, 755.

Vogel, T. (1965) Systemes Evolutifs. Gautier-Villars, Paris.

Volterra, V. (1927) Mem. del R. Com. Talass. Ital. 131; Also in Opere Matematiche, Vol. 5, Rome 1962.

Volterra, V. (1934) C.R. Acad. Sci. Paris 194, 1684.

Wangersky, P.J. and W.J. Cunningham (1957) Ecology 38, 136.

Wheldon, T.E., J. Kirk and H.M. Finlay (1974) Blood 43, 379.

White, B.S. (1977) S.I.A.M.J. App. Math. 32, 666.

Wickramasinghe, S. (1975) Human Bone Marrow. Blackwell, Oxford.

INDEX